Advanced Design Cultures

Manuela Celi
Editor

Advanced Design Cultures

Long-Term Perspective and Continuous Innovation

Editor
Manuela Celi
Dipartimento di Design
Politecnico di Milano
Milano
Italy

This book, completely revised and partly rewritten, was originally published in Italian by Mc Graw Hill under the title *AdvancedDesign. Visioni percorsi e strumenti per predisporsi all'innovazione continua.* Milano, Italy, 2010.

ISBN 978-3-319-08601-9 ISBN 978-3-319-08602-6 (eBook)
DOI 10.1007/978-3-319-08602-6

Library of Congress Control Number: 2014948768

Springer Cham Heidelberg New York Dordrecht London

Printed on acid-free paper

Springer is part of Springer Science+Business Media (www.springer.com)

Preface

Nowadays, the term Design is more extensively used than ever before with reference to a 'galaxy' of meanings or better, to use the words of Peter Lunenfeld, to the *Design Cluster*.[1] In this very broad context, design research (for or with or through design) has acquired increasingly defined value with time: describing practices, identifying methods, establishing invariants or perfecting tools are activities which have expressed the numerous approaches to this discipline.

The problems, environments, methods, tools and case histories presented in this publication describe a specific area of the *Design Cluster*, part of which is still undefined like a foggy cloud, which is extremely interesting because it is transversal to the thematic-applicative areas in which we are used to subdivide the discipline. *AdvanceDesign* has never been told, read or theorized but it is a consolidated practice in product design. Born in the car sector and then extended to other production domains it is undoubtedly linked to the future dimension of design but also to its specific innovation methods. If the initial phases of the process of innovation are internationally recognized as fundamental in this historic moment through the emerging of themes regarding the Front End of Innovation, then we are deeply convinced that the area of Design, our discipline, has a lot to say. Creating *AdvanceDesign* means radically innovating but also innovating through unexplored routes through the involvement of users or imagining plausible and unexpected scenarios.

AdvanceDesign, *Envisioning product and process form* is the name of a research group that between 2009 and 2012 worked first on this theme and then produced this publication. A name that has an identity-related value and that encapsulates the research team's intentions in the claim.

[1] In the preface of the publication by Brenda Laurel on Design Research, Lanfield poetically compares the current extent of the disciplinary area of design to that which for an astronomer is a combination of galaxies, or a *cluster*. P. Lanfield, The Design Cluster, preface to Design Research, Brenda Laurel ed. MIT Press Cambridge Massachusetts 2003, pp.10–15.

Envisioning: imagining, visualizing and drawing are activities that are closely linked to design action. The type of design know-how is characterized by visual knowledge that distinguishes it from other forms of know-how.

Product and process: the new complexities of the product, its new qualities and the ways they are expressed and perceived, but also their social, economic and environmental impacts lead us not only to consider the product but also the process that generates it, as the design subject.

Form: form not for its own ends, but as the image of transformations; form as the capacity to visualize future possibilities and render them tangible; form as an integral part of the innovation process.

AdvanceDesign is not just highly developed design, but also design which anticipates, which sees before others. The analytical context in which this research is placed is made of a combination of two faces: on one side, the various theoretic approaches, methods and tools that characterize it and, on the other, the forms assumed in the various fields of application narrated with the spirit of *reflective practice.*[2] Some of us who, right from the start, decided to devote themselves to the historical and theoretic reading of the phenomenon of *AdvanceDesign* produced brief documents, presentations and sometimes interpretative schemes, and subjected their reflections to the group, looking for a reading key that could be shared. These contributions were gradually formalized and form the first part of the publication where the context in which this research was born is clarified, along with the definitions, the historical dimension, tools and approaches that have characterized *AdvanceDesign*. The second part of the publication contains a series of contributions that look at specific environments in which *AdvanceDesign* has found fertile ground. From cars to electrical appliances and lighting design, from business networks to technology parks and the creative development of innovative materials: different roads to explore the dynamics between research and design.

In the ambivalence of the meaning of *AdvanceDesign* the two souls of this discipline are expressed. On one hand, a hard nucleus linked to the tangible dimension, where design is a technical environment focused on the development of the product and on production materials and technologies. On the other hand, a soft, more ethereal essence, linked to intangibility, where design becomes the encoder of social, cultural and emotional needs and produces new forms and meanings through the creative dimension.

This collection of essays aims to trace the contours for *AdvanceDesign*, identifying the ingredients and invariants that have emerged from the historical reading and, at the same time, generates a series of openings towards emerging research themes.

---

[2] Shön D.A., *The reflective practitioner*, Basic Book, New York, 1983. Trad it. *Il professionista riflessivo*, Dedalo, Bari, 1993.

Contents

Author Biography

Manuela Celi, Ph.D. She holds Ph.D. in Industrial Design and she is currently an Assistant Professor at the Dipartimento di Design of the Politecnico di Milano. From 1999 to 2009, she has cooperated with the School of Design of Politecnico di Milano in designing educational programs, coordinating orienteering projects, and tutoring activity for foreign students or students on stage. She teaches Industrial design at the undergraduate degree course in Furniture Design and Product Design since 2006, and from 2014 she is teaching also at the graduate course of Design for Product Innovation. Her earliest research interests were focused on the forms of knowledge related to design, on their use and translation into skills within the learning systems. Her Ph.D. focuses around design knowledge and its forms of learning converge on the activities of *Metadesign* with the purpose to provide an approach to design knowledge, to learn how to learn, to develop metacognitive skills, to acquire autonomy in coding and decoding information. From 2009 to 2011, she has been the Secretary of the Advanced Design research group and she has deepened her studies in this area, being interested in the methodological approaches, process design, and participatory design. Her most recent research interests are concentrating on advance design and on design process to foster innovation as long-term strategy. Together with F. Celaschi and L. Mata Garcia, she published the article The Extended Value of Design: an Advanced Design Perspective on the DESIGN MANAGEMENT JOURNAL. From 2010, she is one of the Coordinators of the Humanities and Design Lab. The Humanities and Design Lab is a scientific trans-disciplinary research lab that investigates the relationships between the design disciplines and human and social sciences.

Part I
Introduction, Theorical Aspects, Interpretative Keys

Chapter 1
AdvanceDesign Points of View

Flaviano Celaschi

1.1 The Context

In the spring of 2010, I coordinated the foundation of a team of researchers and lecturers who, at the Dipartimento di Design of the Politecnico di Milano, decided to focus their scientific and didactic attention on the design of processes. In particular, it was our aim to study design-driven innovation and the transformations it was facing and in a great part was already subjected to, in order to respond to the changes of an increasingly complex organizational context, one which is economically more globalized and socially and culturally weaker and increasingly interdependent.

With this research group, we tried to condense the common research front into a field of process design that we call Advanced Design[1], reviving and integrating two terms that in the past decades have had their ups and downs. 'Advanced' is a word that was used in the practice of innovation within corporate contexts,[2]

[1] Cfr. Manuela Celi in the second chapter of this book, describes a part of the journey we shared to identify her field of research and try to name it.

[2] Cfr. Celi, the author mentions in her contribution a large number of companies that, starting from the 1980s, have introduced in their institutionalized innovation processes structures and enterprises referred to as 'advanced centre' or 'advanced design innovation', etc.

F. Celaschi (✉)
Alma Mater Studiorum Università di Bologna, Largo Risorgimento, 2, Bologna, Italy
e-mail: flaviano.celaschi@unibo.it

M. Celi (ed.), *Advanced Design Cultures*, DOI 10.1007/978-3-319-08602-6_1

but substantially scarcely theorized or treated by the international academic design research.[3] We can state that Advanced Design has set its roots in the practical field of enterprises and of government research organizations. The word advanced has been exploited in marketing and advertising and for many years it was connoted by a propagandistic meaning which was kept well away from the world of theoretical research. Only in more recent years has it attracted the attention of a discrete number of experts and researchers in design-driven innovation processes. This word qualifies the design through an adjective which associates the meaning to time, intended as long term. But in a progressive vision of the production and social world, it also means something more advanced in terms of performance than before. Here, in general, advanced design studies a phenomenon that tends to overcome the current vision and look beyond; to move into unexplored territories, towards an 'advanced future'.[4] But the emphasis on the time factor is not the only characterizing factor or point of view that we have given to the term 'advanced'; in this book we want to explore some of these aspects of advanced design around innovation.

There are some side conditions that define the context in which our observations on advanced design take shape. The first is undoubtedly 'complexity'. The study, experimentation, design, prototyping, introduction on the market and follow-up in time of the processes of constant innovation of new products and new services, with respect to global social and economic conditions, occur within a combination of relations of enormous importance, an integration of factors never seen before. The context of design and the number of interconnected factors that trigger new phenomena has greatly changed, also in contexts far apart from those from which they originate, and of difficult comprehension and management.[5] The project is doubly involved by complexity because nowadays we are more aware of and attracted by the complexity of the products and services we design, and also with the complexity of the processes with which we design them. The designer's laborious job is to materialize and give form/meaning/value/function (Celaschi 2008) to goods and

3 In the same period David Bihanic, at the University of Auvergne - Clermont-Ferrand, France, began his work on Design and Prospection using the term Advanced Design, Roberto Poli, of the University of Trento in Italy, was able to open an institutionalized scientific and academic debate around the Future studies, launching the preparatory work of the subsequent years around the establishment of the first Master in Future Studies at the University of Trento. In the same period, there was an increase in the number of authors who used terms such as future, long-term research, long-term projects, advanced future, within a context, that of research in design, in which, strangely, it was difficult to explicitly use the word 'future'. In this period Norman (2008) and Thackara (2009) introduced the concept of innovation in due time, in the long term.

4 If we attempt an analogy with respect to the English language, we can say that contemporary design works in respect of two different times: a 'progressive present' that has deep roots in the study of what is already present today to make it evolve with respect to the need for change that consumers are expressing, or a 'future' which is definitely projected forward with respect to the present and which in Latin languages is defined as 'futuro anteriore'.

5 On the concept of complexity, chaos, nonlinearity of systems, Lloyd 2006, pp. 159–182; Bocchi and Ceruti 2009; Morin 1983.

Table 1.1 The principal factors that characterize the context in which contemporary design works and the main consequences that affect the design-driven innovation approach

Phenomenon	Reaction
Complexity	Nonlinear systematic vision of phenomena, permanent imbalance of factors, self-organization and adaptability (ability to adapt) of the systems: from the concept of product design, to the concept of product system design, to the concept of design of experience, design of training and self-transformation (Pine and Gilmore 1999)
Acceleration	Innovation crisis as an episodic factor and orientation towards constant innovation (Celaschi 2007)
Mediation	Design as a mediation between knowledge as a direct mediator between needs; direct access to sources of knowledge and experience and crisis of the traditional mediations (Celaschi 2008)
Sustainability	Crisis in marketing and all short-term visions regarding the transformation of goods and services

services which we must always treat as complex and inseparable systems, made of nonlinear factors, in which equilibrium is extemporaneous and inexplicable (or unexplored). The designer is called to give reality to goods for which the 'rate of reality is reduced day after day' (Baudrillard 2009). If the designer in the first part of the twentieth century was well acquainted with every part of the product he was designing and every material the object was made of, the designer in the second part of the twentieth century had to frequently admit the existence of a 'black box', or rather a component of the object about which he knew only the control and interface terminals. The contemporary designer knows only the human reaction to the recipient of a product and almost nothing about the technologies and mechanisms he is shaping, and to formally and functionally manage them he avails himself of countless 'specialist translators'. At the same time, the transformations that the contemporary designer's project produces transcend well beyond the boundaries of the specific object and invade a considerable number of side phenomena (economic, cultural, social, technological, etc.; Table 1.1).

The acceleration of production processes and consumption is another factor that characterizes the context we are studying. The speed with which complex systems interact through propagation of nonlinear phenomena that cannot be simplified or foreseen is an aspect that innovation processes cannot ignore. The first consequence of this acceleration is that the classical design, founded on:

- an initial trigger;
- a defined amount of time and energies;
- a result requested by a buyer in relation to a precise brief, assigned to the designer;
- a linear process (funnelling from chaos to the solution) which starts from the calm, introduces temporary disruption, to then return to a state of calm;

is no longer effective with respect to the needs of a production and reproduction system in order to have a constant and increasing flux of innovation in time.[6] The inevitable answer to this crisis determined by acceleration is rethinking of innovation processes as continuous and characterized by constantly open corridors of investment, by intermediate outcomes that are constantly shelved, pending the development of solutions to adapt to the temporary needs of the market or society that enable to obtain, in time, the desired result with a reduction in risk and economic investment, acceleration of results and expected quality.[7]

A further element that characterizes the context in which advanced design develops as a paradigm of interesting innovation is the insufficiency of each single discipline to represent, in an isolated and exhaustive manner, a combination of competences suitable to sail the ship into port. This is where the designer's potential comes in, as a mediator between competences and, in particular, increasing the cultural and linguistic quality of the designer, the ability to integrate the knowledge of art with that of economy and management, between the humanities and those linked to technologies and engineering (Celaschi 2008).

Finally, the limited resources calls for a quick and radical transformation of the system we call market and of society, with regard to the environmental repercussions, as also social and democratic, and access to the same resources. The designer as a creator of new goods and services has, in this respect, considerable responsibility in sensitizing all actors involved, and a relevant, needed increase in one's competences with regard to the ability to evaluate the systemic consequences of one's actions. A radical reaction in design towards the negative consequences on this front is represented by the development of the habit to evaluate in the long term the consequences of the project action. The long term as an inevitable horizon also with regard to the goods and services destined to be immediately launched into the process of fruition.

Complexity, acceleration, transdisciplinarity of competences and sustainability help us trace the boundaries of a system within which to attempt to operate design-driven innovation. The integration of these four factors determines a problem field which is completely new, full of new problems and just as many relevant opportunities to prove that the project culture can be, today more than ever, a relevant and effective answer to change. This context that I have tried to summarize is characterized by the difficulties experienced by some classical models of production and reproduction that I would like to sum up: crisis of the paradigm of design as a discontinued act that has in

---

[6] We wish to stress here the increasingly less effectiveness of a process of innovation of goods and services based on the temporal dynamics determined by the concept of project, or rather a process with a scheduled beginning and an end, given resources, objectives defined through a brief.

[7] In this case, the meaning of constant innovation is synonymous with 'shelf innovation' or rather the capacity to carry out constant innovation processes (without solution or obtaining finished product, but creation of components for innovation and a platform logic which is expandable and increasable). The term in this case does not stress the idea of incremental innovation versus radical innovation, but refers to the continuity over time of the research and design flux. Cfr. Celaschi and Deserti 2007.

a brief, a beginning, given resources, and an end; crisis of the all-round designer able to represent effectively and efficiently, and at the same time, the demiurge of each of the phases of innovation, from know-how to engineering; crisis of all short-term visions of change and, above all, crisis of the typically marketing-oriented vision that the recipient-consumers know what they want and can therefore guide the production process and offer better and innovative goods and services. Within a logic of guided and controlled sharing and coproducing value, based on open source of knowledge, the contribution of all to innovation and to the creation of value can and must effectively be enhanced: this is at the same time a problem and a great opportunity for contemporary design and, therefore, also advanced design.

1.2 From Project Culture to *AdvanceDesign*

In the following chapter, edited by Manuela Celi, we describe in detail the process of gradual transformation of advanced design within a process of evolution of the scientization of the project. It is as if we were confronted with a four-stage model of the evolutionary journey which can narrate advanced design as the phase of a process oriented towards a specialization of project knowledge but, at the same time, also towards the expansion of the field of observation and hierarchic growth of the importance of the problems faced with respect to traditional problem levels that every organization and enterprise tends to develop while at the same time needing to grow and consolidate.

A process that began with the acknowledgment of the existence of a 'project culture', perfectly identifiable with modernity, intended as a category of the evolution of the mind (Maldonado 1987). The progressive drive of the project together with the efficiency of the classical production models, massive and self-preserving, reveal the competitive advantage existing in the competition between companies, organizations and individuals that are able to program their own transformation and growth using the proactive episodic nature of the project compared to those companies, organizations or individuals that continue to use proto-modern development processes. The origin of project culture is founded on a linear vision of time (like Chronos), on a previously programmed use of resources, on the action made considering the contexts to design as finite systems of autonomous variables and Cartesianically decomposable,[8] on the faith in the existence of the right and univocal answer to man's problems, on the equivalence between mathematical or

[8] A complex problem is constituted by a more or less vast combination of lots of minor problems, more easily solvable: the solution of n problems that make up the problem of higher level that contains them determines the automatic solution of the problem of higher level.

geometric models and reality, on determinism and on the deduction deriving from analysis and on the practically infinite repeatability of processes.[9]

In this period, within the project culture, there develops the hope of giving a scientific connotation to the creative act of design; the 'design methods' stem from the absolute equivalence between project (to project, to plan) and design (from the Latin *designare*, designate or rather give significance to). The first crisis of design methods concerns the impossibility to scientize also the behaviour and the combination of univocal characteristics that qualify each designer.[10] Within the paradigm of the scientization of design, every problem, even if it has only one right answer, will be solved in exactly the same way by any project designer. But in contemporary design every problem that allows for infinite valid solutions requires that every designer tackling the said problem be unique and different and offer a creative solution to the problem (both in terms of process and product). The design methods cause some evident problems to the design, but at the same time they are a phase of the evolution of the discipline that brings together design with applied sciences and admits it, to a certain extent, into academies and universities, partially taking it away from the supremacy of art.

At the end of this 'design methods' phase, there is a flaring up, especially for cognitive process experts, a growing interest in the ability and processes through which the designer solves a problem. It seems difficult to understand how the designer is capable of integrating a technical act with a creative one.[11] This natural ability can be explained through the processes of abduction that lead the designer to use his own identity to observe and create a model of the reality he will have to transform. The designer puts himself into the solution of the problem that he is tackling, and he does so in every phase of the project process: in observing reality, in constructing simplified models of reality, in the phase of transformation of the models, as also in the evaluation of the pros and cons of the transformation proposed, as well as of the transformation into reality of the transformed model. Every phase is creative, every phase could be solved in a different manner by any other designer. Some observers feel it is possible to extract this way of creating something new from the designer culture and the designer's overall training. As if there were a sort of magic formula through which, at a certain point, and predictably, the designer pulled out a weapon, a way of thinking which is transferable to other competences, to other professions. The concept of design thinking begins here, intended as a reduced portion, a synthetic extract of the actions of the designer, who

---

[9] That which has given a satisfactory result will continue to do so also in the future, or rather you do not change a winning horse.

[10] 'Projecting a project, as the etymology suggests, means to produce an idea in a way that it gains its own autonomy and is realized not only by the efforts of its inventor but also of those that are independent of one's ego' (Czeslaw Milosz in Brodskij J., Miłosz M., A conversation between the two great authors, on literature and not only, published in *La Repubblica*, 4 December 2011, p. 52–53).

[11] From which the famous phrase by Giovanni Klaus Koenig: 'the designer is a bat, half bird, half rat'.

according to the supporters of this ideology, anyone can understand and apply, independent of training and approach.

This phase is steeped in emphasis on creativity,[12] on lateral thinking, on design as an exportable aptitude; rapidly design from practice becomes science, and from science turns into a brand that identifies the carriers of a sensitivity and capacity of autonomous thinking. In this phase, design seems to have become the panacea of all of society's ills and the solution to all the problems in innovative change. From this phase of apparent fortune and pervasiveness of the concept of design a critical condition is generated, primarily inherent to the designers and experts of creative design processes that do not recognize as separable the cultures of the creative thinking project. Alessandro Deserti and Giulio Ceppi in the essays that follow, analyse this concept according to which design thinking is not exportable from design; or rather the ability to innovate, intended as a sort of resilience of the intellect, a creative aptitude, is not innate and cannot be formed separately from project design competence: you learn to design by designing and at the same time you learn design thinking or rather the modus operandi of the designer in designing a project.

Today, it is no longer possible to confine design to the category of problem-solving. Every phase of the project, from problem finding to the evaluation of the results obtained, are infused into the project culture and creativity intended as the designer's aptitude to put himself in the observation, research, solution, transformation and in the final evaluation of the design process (Celaschi 2011; Celaschi et al. 2011). Design tackles problems that do not have just one exact answer, but infinite possible answers in competition. Design controls processes and products in which it must decide the balance between sense/value/form/function; processes and products in which the concept of time is not that of 'chronos', absolute time measurable in intervals and linear in passage, but rather that of kairos[13]; design operates in contexts that are not just the industrial one, it

[12] From Bateson's first studies on self-poietic processes and on the self-organization ability of organic systems, we get to Shon's studies in just over a decade. 'In 1993, Schon suggests a learning concept that finds its fundaments in the concept of reflection in the course of action. Our knowledge lies in the same action. He investigates the processes of knowledge and learning occurring during the course of the action (professional practice) reaching the definition of an action of reflective nature which, starting from the uncertainty and anxiety connected with it, can become itself the generator of a new knowledge. The author reaches the assumption that the solution of problems implies first of all the process of definition of the problem, through which the action to take is decided, the objectives to set and the means to choose'. Cfr. www.formazione-esperenziale.it. It is the act of designing and the project dialogue between the authors of the process that lead to the construction of other possible worlds. There is no creative thinking which is separable from these processes and applicable to other professions (Sclavi 2003). Other authors describe the concept of design thinking as not extractable from the project cultures towards other practices, like for example towards management, without having an adequate creative aptitude formed, in fact, through designing.

[13] 'Kairos as a time in between, a moment of an indeterminate period of time in which 'something' special happens. That which is special depends on who is using that term' in Zaccaria Ruggiu 2006.

moves within a holistic vision of the context in which in designs, it practices rules and at the same time breaks them (Celaschi and Formia 2012).

It is within this vision of design, that I call 'design cultures', using the plural, that it starts to make sense to talk about advanced design as a portion of the field of investigation around innovation processes that are able to offer new and interesting points of view.

1.3 Three Moments in *AdvanceDesign*

The first systematic use of the concept of advanced design, from the period immediately after the second World War until the end of the technological and project supremacy of western culture, that coincides with the landing on the moon and therefore with the American Apollo project, is strictly correlated to the positive and optimistic dimension of industry and the mass consumption that derives from it. The ever-more capillary circulation of goods and services offers huge economic resources to the traditional industry that drives ambitions and competition amongst multinational industrial giants. This drive is at the same time a quantitative and qualitative one, and one of the frontiers to conquer more obsessively concerns of science fiction dimension produced by western literature, comics and cinematography blandishing us increasingly towards the dream as a materializable dimension.

It is within this context that the leading international expos and trade fairs become exceptional venues in which to stage the competition between enterprises and the capacity to aim beyond the feasible and lead us to lose contact with reality; products are created that are no longer the logical consequence of an incremental innovation with respect to models previously produced, but are the moment of the 'dream car' and the 'concept product', futuristic goods that are not conceived for material consumption but for conceptual, idealistic consumption. They are goods that are consumed because they represent future visions which can confer an oneiric drive to the far more normal goods produced and distributed by those same enterprises for the real market. In this first phase of advanced design, we in fact give this definition to company divisions that are mounted and formed to respond to this need to create a communication value and at the same time a long-term vision of the designing abilities of a brand. Frequently in these goods, the communication value largely outweighs the real performance value of the pilot product conceived for the evolution of a 'species': aerodynamic properties, lightness, automatism of systems, transparency, etc., are keywords that must be materialized in these concepts that pass directly from the project centres to design museums without actually having an impact on productivity in the following years. From the 1990s of the past century on the pervasiveness of the ICT and the evident and massive consequences in everyday life to their introduction on the market produce another effect with respect to advanced design, which becomes the answer that the world of production offers designers to integrate material values like form and function with intangible values like that which the digital revolution has brought. Designers are too few to

Table 1.2 The three phases of *AdvanceDesign*

Evolutionary drive in advanced design	Focus
Design plus dream	In the West, liberal and free trade optimism of the late post-war period is conditioned by the American Dream (consuming generates happiness) and by the idea of limitless resources. In this context, the barriers of sustainability are rapidly crashed to the extent of considering this dream as a reachable objective towards which to direct project efforts. It is like a drive to orient the rationality of a classical production system towards the irrationality of impossible, but imaginable worlds. An example of products of this phase are the 'dream car' or 'concept vehicle' flaunted in popular universal expos
Design plus technology	The evolution of ICT and the advent of the computer, the mobile phone, Internet, generate the disorientation and crisis of the traditional processes of materialization that design had followed until then. The designer no longer controls the technologies and the knowledge that determines the operation of the products and services but is just limited to managing the interfaces. Examples of these kinds of products is digital graphic design and the synthesis of the form of personal computers and the world of Apple, standard devices but customizable at the same time
Design plus future studies	The pervasiveness of design as a culture relieves the designer from the obligation to create industrially produced objects and leads him into a more expressive dimension of his social and economic role. At the same time, future studies become a branch of Humanities and also offers design a wealth of instruments to manipulate the advanced future as the project's horizon: pre-emption becomes a competitive advantage and a language through which to involve the consumer in the coproduction of value. An example of these products is the creative development centres established by companies and organizations, experimental urban planning, smart cities

understand and manage the quality of products and services about which they now know only a small part of their technical- and performance-related characteristics. We must give goods in the digital revolution a form and above all an interface that is able to pursue the progressive and speedy rarefaction of the dimensions and weight of things.

In this phase, advanced centres of enterprises and companies are places stacked with computers with which to attempt to prove that calculation power and quality of life are two correlated variables. Advanced and digital become synonyms and design, lacking the material dimension of goods, approaches the service project and that of the product content with increasing intensity (Table 1.2).

The third phase, which we can define contemporary to advanced design, is one which generates our interest in this approach and at the same time the fortune and usefulness is revived in this term, which we shall try to explain in this book.

The strongest drive that makes this approach evolve comes from humanity studies, and in particular the so-called future studies, a branch of contemporary sociology that studies, in a scientific and systematic fashion, the problem of anticipation, offering a repertoire of methodologies and approaches that enable to study the future as a phenomenon and to build courses of 'systematic construction' of sceneries and products sciences contained therein.

Complexity, acceleration, integration of competences and sustainability support the inevitable long-term vision as the true challenge to approach in order to generate value in respect of shared values. Shared anticipation of the future is no longer a demonstration of public economic power, but an inevitable necessity without which there would be a development and social regression. Broadening the vision becomes therefore a broadening in space (globalization) and in time (long term) and it seems that only in this dilated space/time context would it make sense to design with the aim to support a growth that is no longer associated to the unlimited increase in consumption, but with an increasingly deep sense of the term development, which is associated with the quality of life, the well-being of people, the increase in sharing of resources and with the capacity to tackle, in a collaborative and systematic manner, the big problems of the earth (climate change, hunger, poverty, demographic increase, pollution, …)

1.4 The Present Field of Research

Today, we define as *AdvanceDesign* (ADD), an articulated combination of design processes that attempt to give shape to products and services destined in the advanced future. Processes destined to produce goods for very complex contexts, through the involvement of extended groups of designers; projects which are often requested by a specific buyer; projects that often tackle situations that have no connection to the present, nor production sectors of reference because they are extremely innovative and unusual with respect to the normal panorama of goods produced for immediate consumption.

ADD deals with projects that often still do not have a market of destination, a production sector of reference, or a customer. These are projects that typically do not have just one creative author but are often targeted at other designers rather than at the final consumer market.

ADD is in the front-end of design-driven innovation, and therefore needs constant theoretical reflection, rooted in practice, for creating its own instruments and courses that can be followed to provide the future with original forms. ADD's goal is to transform every discovery, every new garden of knowledge, every invention into constant innovation, not only adapting them to the expectations and needs of the production system and of the end-user (the market), but contributing to create new products, new production processes, new users and new markets distributing innovation.

The general definition of design as a process of transformation of reality based on the observation of the phenomenon, on the construction of simplified models of this reality, on the manipulation of the simplified model, on the evaluation of the pros and cons of this manipulation with respect to the previous reality, and finally on the transformation into reality of the manipulated model, perfectly fits also ADD.

The main fields of knowledge that determine the formation of the profile of a designer (humanities, art, technology, economy and management) are the same that form an ADD operator: in the project result of the ADD, the form/function/meaning/value factors are integrated, as in the typical design.

As much as many instruments normally used by the designer in his design research are effective instruments for working in the ADD field (tools like *scenario building*, *mapping*, *modelling*, *conceptualization*, etc.), the ADD operator must pay particular attention to the self-determination of his tools and to the original and constant redesign of the same design processes. Particular and extreme situations strongly call for the need to develop delicate tools that are adopted on an experimental basis, often with the possibility to subsequently transfer their use to 'ordinary' contexts.

ADD is often characterized by the absence of an actual buyer of the project, which leads to models in which the project activity also consists in obtaining the resources needed for the same project.

The logics of *feasibility* or *producibility*, normally adopted in the assessment of the project's progress, are issues in the ADD field, in which the absence of precedents in the market with respect to the project designed makes it difficult to make decisions on the projects developed.

The activation of collective design processes, that involve a significant number of experts that operate in different fields, has led to the fact that in ADD there are often personality clashes deriving from the different visions and creative personalities that have to be managed.

The time frames of the ADD projects are usually very long, with the consequence that the fields of application can suffer considerable disruption—from the project launch to its design and to its finalization—that require adaptation and adjustment of the project while in progress.

ADD often skips the traditional forms of mediation between interest typical of the design process: marketing, accounting, distribution, etc., are of not much use to ADD projects, that must, instead, often mediate between the producer of knowledge (the scientist), the potential production system (the enterprise) and the potential consumer (the market), that frequently are not able to understand each other's language.

1.5 Four Areas of *AdvanceDesign*

In the definition which I refer to there are four main directions the ADD research can follow, very different from one another but coherent in terms of ability to drive constant innovation through the culture of design.

1.5.1 Design and Time

The most characteristic dimension of ADD is that associated with the time factor. We speak of ADD when the time horizon of the project is shifted significantly forward. In this situation, it is typically necessary to elaborate solutions that do not incrementally improve what already exists, but instead indicate directions and trajectories for unprecedented innovation and a particularly original one (*disruptive innovation*). This is the significance that has been attributed to the term ADD by the first production sectors that have developed and implemented operative centres for designing ADD solutions: *automotive* (concept cars, dream cars, advanced prototypes …); consumer electronics, etc.

Design research concerning issues of anticipation and materialization and narration are part of the work carried out by these centres. Projects aim at the construction of scenarios regarding relatively distant futures and at the development of solutions destined to explore the future of various different sectors of production of services and goods.

1.5.2 Extreme Design

There are design-driven innovation processes that pursue innovation beyond the production sector of reference or in places that are geographically and culturally very distant from those to whom the results of the project are destined for. An 'exotic' direction of the research that sees materialization of something new from the courageous transfer of innovation through connections and logics which are usually and normally incompatible, used often in the name of a research of ethno-anthropological nature with strong integrations between the humanities and design.

Within this scope, we can appreciate the results of project research that have developed tangible outcomes and produced products or services which have entered into production starting from the exploration contexts which are very different and apart from those of the recipient of the project. These can be activities that investigate the past to identify potential comebacks, cycles and periodic resurfacing of tendencies, or that observe preindustrial behaviour to determine post-industrial stimuli to daily use and consumption. These can be investigations on extreme situations (*extreme design approach*) in which experiments are carried out to find stimuli that in the normal context are difficult to isolate. Here, we can speak of transfer of innovation from contexts and conditions and users that are hardly associated with those in which the design is called to plan.

1.5.3 Shelf Innovation and Research's Tools Design

ADD is also often characterized as a BtoB (*business to business*) in the sense that its recipients are operators, designers, innovators, entrepreneurs, etc., who can utilize the innovations designed as instruments and semi-finished products useful for acceleration and strengthening of the innovation capacity, actually materializing a proper design-driven innovation supply chain.

The contemporary designer operates in mature contexts that are crowded with intermediate operators, actors that develop only one part of the complex innovation process. In all these cases, we can observe the growth of profiles of designers who work at the source of the traditional design processes of semi-finished products or of components or project stimulation instruments (like, for example trend reports, future scenarios, paradigms of user centred research, etc.).

Within this field, we can consider project research that concerns the creation of instruments and processes for triggering, fertilizing, systemizing and aiding the development of an actual product or service. These are projects whose users are other designers, creative operators, entrepreneurs, innovation operators who need to reach elevated performance levels in a short time and with such levels of specialist competence that they are not compatible with long-term research. We could therefore define this area of ADD as the area of design of the process of design.

1.5.4 Design Without Market

There is a type of ADD which we can define as 'without market' because it is characterized by the absence of an industrial demand in the traditional sense of the term, or rather of a contract (with an enterprise) that will implement the output of the project, funding the operators involved according to the provisions of the project's traditional market. In this direction, the absence of a specific demand or need gives greater importance to the actions of the designer as a self-producer of his own brief and of the requirements that define the project. This area can generate research results and projects that stem from a social appeal, from the involvement of communities of practices, from creative communities, from the triggering of a bottom-up need for design, from the autonomous observation of needs and wishes that have not been filtered by the market nor by entrepreneurs that commission their resolution and transformation in goods or services that can be sold and bought. In this direction, we also find projects and project researches that promote the creation of briefs and scenarios that imply the need to create new organizational models or start-up enterprises designed around the actual project because at that moment there is none that can materialize this product or service.

1.6 Where Do We Start?

The birth of advanced design originates from the practice of entrepreneurial action and evolves rapidly in large public organizations that are called to manage large products on a medium–long term that need an anticipatory structure in support of changes of great social, economic or military relevance and changes in the conditions of man's life on earth. The journey of advanced design seems to encounter a setback when the contingency of the marketing centred projects becomes indispensable to drive an industry in decline due to the change in global scenarios. In these conditions, many organizations operate on a short term and abandon long-term projects and strategies that surpass, in terms of time, the dimension of the 'balance of the company's accounts'.

In this period, which sees the advance of the post-industrial culture of optimizing the existent and the connection between nodes of the network as a source for generating and sharing knowledge, the science of the project begins to worry about matters connected to the advanced future. It encounters a ponderous current of knowledge generated by the future studies included in contemporary sociology and begins to theorize processes that until then had been more practiced than studied. It is the origin of a new vigorous phase of advanced design that through the definition that in this essay we have tried to isolate, it is increasingly necessary to allow us to have a horizon of reference within which to continue (more or less constricted by critical factors) to animate guiding visions and to design processes which can look at the long term. The project that becomes design and constant innovation starts to feed conceptually and pragmatically on the instruments and processes which can effectively develop the future (Fig. 1.1).

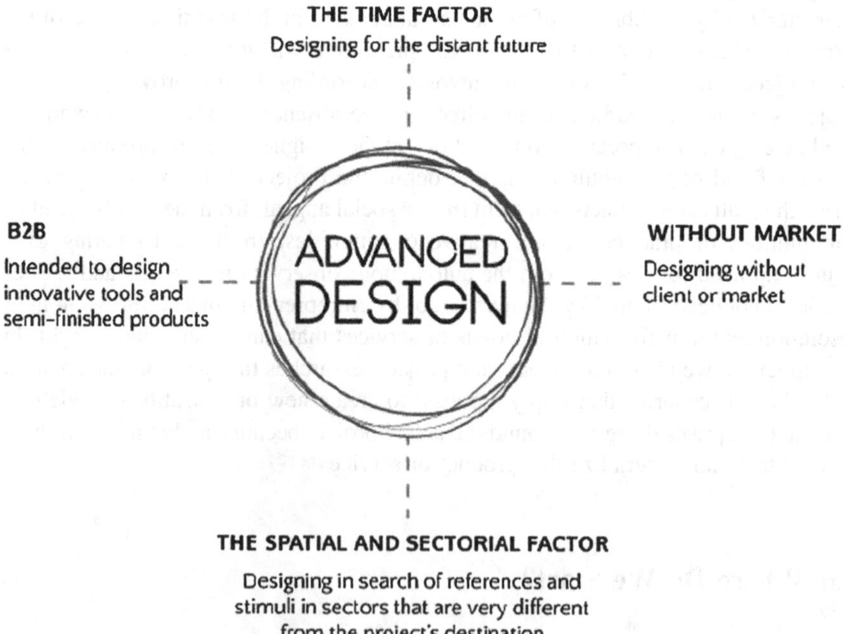
THE TIME FACTOR
Designing for the distant future
B2B
Intended to design innovative tools and semi-finished products
ADVANCED DESIGN
WITHOUT MARKET
Designing without client or market
THE SPATIAL AND SECTORIAL FACTOR
Designing in search of references and stimuli in sectors that are very different from the project's destination

Fig. 1.1 Summary of the directions that *AdvanceDesign* follows to drive innovation in organizations and enterprises

References

Baudrillard, J.: La scomparsa della realtà (The Disappearance of Reality). Fausto Lupetti editore, Milano (2009)

Bocchi, G., Ceruti, M. (eds.): La sfida della complessità (The Challenge of Complexity). Franco Angeli, Milano (2009)

Celaschi, F., Deserti, A., Design e innovazione. Carocci editore, Roma (2007)

Celaschi, F.: Il design come meditore tra saperi (Design as Mediator Between Competences). In: Germak, C. (ed.) L'uomo al centro del progetto (Man at the Centre of the Project). Allemandi, Torino (2008)

Celaschi, F.: Advanced design processes in some case studies from the contemporary art system. Strateg. Design Res. J. **4**(1), 1–4 (2011)

Celaschi, F., Deserti, A.: Design e innovazione (Design and Innovation). Carocci, Roma (2007)

Celaschi, F., Formia, E.: Education for design processes: the influence of latin cultures and contemporary problems in production systems. In: Formia, E. (ed.) Innovation in Design Education, pp. 9–18. Allemandi, Torino (2012)

Celaschi, F., Celi, M., Mata García, L.: The extended value of design: an advanced design perspective. Design Manag. J. **6**(1), 6–15 (2011)

Lloyd, S.: Il programma dell'universo (The Program of the Universe). Einaudi, Torino (2006)

Maldonado, T.: Il futuro della modernità (The Future of Modernity). Feltrinelli, Milano (1987)

Morin, E.: Introduzione al pensiero complesso (Introduction to Complex Thinking). Sperling & Kupfer, Milano (1983)

Norman, D.: Il design del futuro (The Design of Future Things). Apogeo, Milano (2008)

Pine, B.J., Gilmore, J.H.: The Experience Economy. Harvard Business School Press, Boston (1999)

Sclavi, M.: L'arte di ascoltare e altri mondi possibili (The Art of Listening and Other Possible Worlds). Mondadori, Milano (2003)

Thackara, J.: Clean Growth: From Mindless Development to Design Mindfulness, Innovation. The Robert Gordon University, Aberdeen (2009)

Zaccaria Ruggiu, A.: Le forme del tempo, Aion Chronos, Kairo (The Forms of Time: Aion, Chronos, Kairo). Il Poligrafo, Padova (2006)

References

[illegible] ... Design ... [illegible]

[illegible] ... Design Issues ... [illegible]

[illegible] ... Milano ... [illegible]

[illegible] ... Politecnico ... [illegible]

Chapter 2
Preliminary Studies on *AdvanceDesign*

Manuela Celi

> *The Greeks had two words to mean life and existence. Life is the fact that we move in the world, are in the world and exist in the world. Existence is the form taken on by life: given the external conditions, political conditions, traditional and cultural conditions and so on. I'm not interested in life at all, I'm interested in existence... As a designer, as an architect and also as an intellectual I do this job: thinking about what existence can be, what it will be or what it has been.*
>
> Ettore Sottsass Jr.

Undertaking an introductive and simplified approach to the study of *AdvanceDesign*, means describing the problematic sphere, talking about the modalities and methods that have characterized it so far, with the aim of identifying new design spaces and innovative research keys.

The research activity evolves between these two tensions: projects and theories, practices and reflections: what does it mean to perform research into *AdvanceDesign*?

The design sphere, constantly searching for legitimization, rose up as a discipline about 60 years ago. In the 1960s, numerous authors, including John Christopher Jones (Jones and Thornley 1963), Alexander (1964) and Gregory (1966), were concerned with identifying a design method capable of transforming design activities into a new disciplinary sphere. The political situation and the economic-social situation after World War II created fertile ground for the development of design. The conference on design methods held in London in 1962, the consequent foundation of the Design Research Society, *operational research methods* and *decision-making* techniques which emerged after WWII, the birth of the first computers and programming techniques represent the basis of the application of scientific methods to design. The English school is rich of contributions to *design as science* through books on the encoding of the design process and on *problem-solving* (Asimov 1962; Simon 1969). However, a decade later, having established

M. Celi (✉)
Dipartimento di Design, Politecnico di Milano, Via Durando 38A,
20158 Milano, Italy
e-mail: manuela.celi@polimi.it

M. Celi (ed.), *Advanced Design Cultures*, DOI 10.1007/978-3-319-08602-6_2

the scarce applicative success of these methodologies, some of these same authors stood back from the excessive encoding of design. Alexander and Jones in particular recognize that the language of machines, behaviourism and the ongoing attempt to bridle design in a logical frame do not represent a useful contribution to the interpretation of design.[1] The rigidity of the first models was somehow bypassed when Horst Rittel, classifying them as first generation models, opened the possibility to develop new methods no longer based upon the optimization and omnipotence of the designer, but on the capacity to recognize satisfactory and concerting solutions with the "proprietors" of the problem (commissioners, customers, users, communities, etc.) (Cross 2007).

Since then, the extensive debate on the nature of research into the design sphere rotates around the keys to reading and around the interpretations: search *for* design, research *into* design, research *through* design, design issues or design studies, design as a discipline.

At the beginning of the 1980s, Bruce Archer and Nigel Cross traced the boundaries of the *designerly ways of knowing*, but it was particularly with Donald Shön (1993) that the practice of design found an interesting interpretation as a 'reflective practice'. Opposing the conception of design as a science, Schön proposed an epistemology of the professional practice through a meticulous analysis of the behaviour of certain 'design professionals'.[2] He explores the ways of viewing design, the roles of prototypes and structures, and the potential of tools to understand how to generate and encode knowledge that is triggered only by practicing the profession in which the professionals themselves do not know how to describe. There is no likelihood of a sublimation of practical skill, but an encouragement to solve the split between 'strong knowledge' (of science and knowledge) and 'weak knowledge' (of artistic ability, practice and mere opinion). Practice, as reiterated by certain authors in the field of social-pedagogical reflection, is not the moment in which 'theory' is applied, but another investigative-operative process, with its own specifics and an original value.

There is also another typical dimension of the design project, which requires to be explored: the fuzzy one. Creativity, casualness, an irrational component, a visionary approach, are all terms which we trace back to that undefined dimension of the design activity, being activities aimed at the future. In his studies, Cross (2007) indicates a series of significant testimonies relating to several well known design methodologists: Jones, who moves away from the scientific approach, stating 'I reacted against design methods. I dislike the machine language, the behaviourism, the continual attempt to fix the whole of life into a logical framework'; Alexander who, in defining the nature of the discipline, admits 'Scientists try to identify the components of existing structures, designers try to shape the

[1] In the 1970s, both authors dissociated or moved away from an excessively stiff reading of the activity and methods of design. Cf. Cross (2007), pp. 1–4.

[2] Schön defines 'design professions' as architecture, engineering, industrial design and the design of software, but analyzes these professions as being examples of the design nature present in every other kind of practical activity (Shön 1993, pp. 16–17).

components of new structures'; and Gregory, who adopts the same line of thought, declaring 'The scientific method is a pattern of problem-solving behaviour employed in finding out the nature of what exists, whereas the design method is a pattern of behaviour employed in inventing things…which do not yet exist'.

The Dewinian roots of Shön's (1993) work, the distancing from the visions of behaviourism, opening towards a second generation of methodological readings have led, in more recent years, to the consideration of design through its own terms, measuring it using its own values, studying its rigorous nature, which is also contaminated by uncontrollable factors, encoding it through the reflexive practice.

What has been the historical position of *AdvanceDesign* in this evolutionary process? What specific contribution can it have to the growth of our disciplinary sphere?

AdvanceDesign in design culture has never been scientifically investigated, but it has been quite widely practiced. In the history of tangible culture and the history of design, the term has been used to identify 'risky operations', projects which have evolved in merchandising spheres which are not always similar.

The decision to indicate *AdvanceDesign* as a double term contracted into a single word arises from the ambiguous use of two terms historically used indifferently, which convey origin to very different meanings: *advanced design* and *advance design*. The debate on *AdvanceDesign* and on its nature starts from the term through the various linguistic and phenomenological meanings acquired in time, which identify it, depending on the cases and spheres, with quite different connotations.

2.1 The Terms

2.1.1 AdvanceDesign as Highly Developed

advanced [-st][3]
adj.
1. at a higher level in training
or knowledge or skill
2. comparatively late in a
course of development
3. ahead in development;
complex or intricate

This is undoubtedly the most frequently used and commonly understood meaning; it means *superior*, *evolved*, *advanced* as the opposite to elementary. In this sense, *advanced design* occurs particularly in relation to detailed projects and,

[3] For all the definitions of the following terms, the following references were used: Picchi (1999); wordereference.com; Cambridge Advanced Learner Dictionary (www.dictionary.cambridge.org).

when linked to *engineering*, is synonymous with a stage of advanced design, considered as mature and complete, or is an indicator of a subsequent version of a previous product, referring to ideas of progress or evolution. In the Dictionary of Collocations, *advanced* is associated with *technically, technologically.*

In 1979, the Olivetti Advanced Technology Center (ATC), was established in Cupertino in the United States, two blocks from the Apple headquarters. It is here that *advanced design* operations were born: *advanced technology* combined with Italian design, giving life, through the cooperation with the headquarters in Ivrea, to the world's first electronic typewriter, the ET101, to the first European personal computer, the Olivetti M20, and later the M24, the computer which was hugely successful, thanks to the partnership with AT&T.

In 1984, NASA together with the University Space Research Association, launched the Advanced Design Program: a special, intensive project conceived to increase the design component in the undergraduate programmes of the engineering schools. The aims of the project implicate the need to strengthen technological competitiveness, increase the creativity of NASA projects, produce an efficacious and effective cooperation between the NASA centres and universities involved in the project, and expand the basic resources for aerospace design. The project had a duration of over 10 years, and involved 43 universities, extending to the aeronautical sector, directly involving the students with companies and projects directed by NASA itself (Johnson and Rumbaugh 1998).

Jodie Forlizzi of the Carnagie Mellon University, specialist in Human Computer Interaction (HCI), proposes a course on Advanced Design Methods in which she expands the fulcrum of the discipline from the themes of *usability* linked to relations between user and system, to the way in which the system itself is placed socially and culturally. The approach to the discipline moves towards qualitative research, towards ethnographic methods to understand the situations and contexts in which technology is used. Forlizzi's course uses a series of specific frameworks of design research, associates qualitative and visual methods and analyzes the processes aimed at creating and planning a design research (Zimmerman et al. 2007).

2.1.2 AdvanceDesign as Pre-vision

advance [əd'vɑːns]
noun
1. a movement forward;
2. a change for the better; progress in development
verb
1. move forward, also in the metaphorical sense
2. cause to move forward

In this sense, the term takes on the meaning of *advanced*, considered as *movement forward*, *progress*, or *early*. It is a connotation that highlights the correlation between *advance design*—which is more correct without the 'd'—and a future of 'futurable' dimension of the project. Once again as a noun, when preceded by 'in', *advance* acquires the meaning of *preventively*, *before* granting *advance design* the faculty of forecasting future worlds.

The term *AdvanceDesign* is linked indissolubly to the *automotive* sector: first *dream cars* and then *concept cars* represent ways of conceiving the automobile with a glance. The dream car phenomenon developed around the 1950s. Designers of the time, particularly in the United States, experimented new forms and technologies, building unique examples. The purpose of this operation consists in predicting the stylistic and technical developments of cars in the years to come.

One of the very first examples of the use of *advanced design* dates back to this period in time and concerns the Advanced Design Trucks with which Chevrolet began a successful new series capable of revolutionizing the world of pickup trucks and bringing it out of the wake of WWII. The formal innovations, the big improvements to interiors and a revolutionary system of forced air convection represented such unexpected changes that the series is that which clocked up the most sales between the year of its debut, 1947 and 1955. The *dream cars,* which had the most bizarre bodywork and were fitted with futuristic propulsions, stayed on the scene until the end of the 1960s. None of those odd vehicles drove along our roads, yet they were the forerunners of a way of performing research *through* design.[4]

The philosophy that led those experiments is very different from that which led to the conception and construction of a prototype and is the same that moved car manufacturers to design *concept cars*: conceiving visions, imagining future contexts to trace evolutionary paths. Building future scenarios using tangible products is a prerogative of the automobile sector which, having to deal with advanced technological development and, at the same time, slow to adapt and transform infrastructures and industry, tried to pilot trends, by creating futuristic models. The current *concept cars*, almost never destined for production, constitute the ground for experimenting new technologies, proposing avant-garde solutions and new styles, testing public opinion through the events in which they are presented. These products seek solutions to the pressing safety, consumption and performance

---

[4] Research *through* design is the form of research that regards education close up: in this case, design is the vehicle for research and represents a means of communicating the results. The even more controversial term was coined by Frayling (1993/1994). Research *through* design was examined by different authors who, on a case by case basis, defined it as *practice-led research*, *action research* or *project-grounded research*, as defined by Alain Findeli. Despite believing that this form of research is merely a variant of design research with an accent on the theoretic aspects, the author highlights the role of creativity and claims its independence from the other discipline (Findeli 2000).

problems, combining technical innovation with ergonomics, increased comfort and pleasure in driving and show how design represents the fundamental link of all the components.

It is from this tradition that the most heterogeneous and modern design centres have emerged over the years.

The Nokia advanced design Team, for example is constantly working on the production of new scenarios through a concept design oriented to medium-term development. For instance Homegrown: a long-term research project aimed at developing sustainable products. The team has explored specific environmental and social issues, including recycling, energy and extension of accessibility to mobile technology to fields of users previously uncharted.[5] The design team that develops these concepts usually works over a period of time between 3 and 5 years in the future.

In the same way, but in another sector, Whirlpool has been promoting projects that look at the future but also embrace the company's traditional values, moving through Global Consumer Design, and this has really paid off in terms of image. Since 1998, with the aim of reorganizing the design process in support of platform products, Whirlpool has placed design at the centre of its strategic progress using an extended innovation process that involves human resources at all levels, systematic use of ethnographic research and the *user-centre design approach.*[6]

Technology parks as well as design agencies, networks born with the aim of producing lines of continuous innovation are just some of the examples of *AdvanceDesign* that we will see analyzed in the phenomenological depth study of this text.

---

[5] The project is carried out by the same team that created Remade, a concept shown for the first time at this year's Mobile World Congress, which explores the subject of the use of recycled materials to create mobile devices in the future. During the event, in the Nokia design studio in London, the team showed some of the other concepts on which it is working for the first time. These are:

• Zero Waste Charger concept, which explores ways of reducing energy waste when battery chargers are detached from a mobile device but remain plugged into the electricity socket;

• People First concept, which involves three universal ideas that belong to the way people conceive communication—time, lists and contacts—to inspire and examine new ideas of user interfaces;

• Wears in, not out concept: while more and more services become available on our mobile devices, the project explores how, in the future, users could potentially upgrade their devices digitally and not physically, giving people an additional choice as to how to use their mobile phones. NokiaPressServices, April 29, 2008, www.pressbulletinboard.nokia.com/2008/04/29/homegrown—new-design-thinking-on-sustainability.

[6] In 2007, the Design Council was invited to perform research into the "Design Process" in 11 big companies. Whirlpool was one of these and the results were published in a series of interventions. Cf. *A Study of the design process. Eleven lessons: managing design in 11 global brands*, www.designcouncil.org.uk/en/About-Design/managingdesign/Eleven-lessons.

2.1.3 AdvanceDesign as an Innovator

advance [əd'vɑ:ns]
noun
2. a change for the better;
progress in development
[…]
4. the act of moving forward
towards a goal
verb
[…] *3.* contribute to the
progress or growth of
advanced [-st]
adj.
[…] *4.* ahead of the times
5. farther along in physical or
mental development

AdvanceDesign is the name of this book because in the summary of this purposely ambiguous name, two souls of the term come together: a more tangible *advanced* nature, linked to evolved and mature projects (Advanced Materials, Advanced Car Design, Advanced Design Studios), and a more ethereal *advance* nature, aimed at the future, where the term takes on the meaning of *advanced,* considered as *movement forward, progress, anticipation* or *pre-vision.*

Therefore, *AdvanceDesign* is a *double-bind* environment, characterized by contrasting aspects, like two faces of the same coin, which show different ways of tackling the project.

In this double connotation, there is a shared root: an *AdvanceDesign* which is synonymous with innovation.

In the long debate that took place when our research group was set up, the term innovation found us unanimous with regard to its centrality with respect to the different paths that had characterized our individual histories. The search for our disciplinary identity began here.

At the time, some people had traced the maps of innovation, studying relations between technique, economy and society, analyzing the characters of technological innovations, their development spheres and the processes of diffusion. Others have investigated the processes close up, while others had examined the involvement of users in the innovation process. In some cases, the study of innovation was sectorial, specific and therefore very detailed, whilst in others it was broader and more disciplinary.

The study of relations between innovation processes and design is a very current issue and we are witnessing a progressive acknowledgement of the role of design in economic growth.

The *Design Council*, for example has carried out a series of studies on the role of design as a strategic tool for maximizing performances and triggering innovative processes even during periods of crisis, also analyzing historical case studies.[7] Not only did Europe establish 2009 as the Year of Creativity, but for the first time, it called for a series of public consultations on how the Union can further support *design-led innovation* with the aim of making an integral part of innovation policies.

In the United States, the US National Design Policy Initiative (2009) was launched to monitor and understand the role of design at internal and global economic level.

A political-economic interest in this disciplinary area emerges, and for many years there has been a tendency to assign a design-related character to high-profile professions. There has been talk of widespread design, of 'a pillaging of the designer's tools' and it seems that anyone who deals with innovations also deals with design, without necessarily being entitled.

There is a disciplinary environment which has moved in this direction more than any other and deserves analysis and reflection: this is the *operation* area, the same area that inspired the first generation *design methods*. At international level, the interest on the topic of innovation, initially aimed at encoding and translating the new product development process (NPD), recently addressed the initial phase of the innovation process described as Front End of Innovation (FEI).

2.2 Front End of Innovation: Descriptive Notes

The first to show an interest in the theme were Smith and Reinertsen (1991) who define as the Fuzzy Front End of Innovation (FFE or FFEI) the very first new product development phase, meaning that period of time spent to devise and mould the idea, even before submitting it to a first official meeting, otherwise known as 'the start date of team alignment'.

The fuzzy nature of this moment in the design process has originated different reading keys. Various authors in search of a linear model have tried to distinguish and subdivide the initial and secondary phases of the FFE. The first are almost always described with vague terms, as a moment of identification, recognition and

---

[7] See the document *Driving Recovery with Design* published by the Design Council in July 2009. The article reports recent data on how design influences the recovery of British companies and proposes a historical reading of the phenomenon to create awareness on the use of specific design policies: 'Design is widely seen as a driver of innovation and growth throughout the UK economy. This will be vital during the recession: new processes, products and services will be a critical part of the UK's recovery. Design can play a key role in these activities, catalyzing and supporting innovation by providing a formal approach to creativity' (www.designcouncil.org.uk/Documents/About%20design/Facts%20and%20figures/DesignCouncilBriefing_04_DrivingRecoveryWithDesign.pdf).

Table 2.1 Differences between FFE and NPD (Koen et al. 1996)

	Fuzzy front end (FFE)	New product development (NPD)
Nature of work	Experimental, often chaotic, 'Eureka' moments. Can schedule work—but not invention	Disciplined and goal-oriented with a project plan
Commercialization date	Unpredictable or uncertain	High degree of certainty
Funding	Variable—in the beginning phases many projects may be 'bootlegged,' while others will need funding to proceed	Budgeted
Revenue expectations	Often uncertain, with a great deal of speculation	Predictable, with increasing certainty, analysis, and documentation as the product release date gets closer
Activity	Individuals and team conducting research to minimize risk and optimize potential	Multifunction product and/or process development team
Measures of progress	Strengthened concepts	Milestone achievement

structuring of opportunities, as exploration and collection of information, as *upfront homework*. The seconds regard the moment of conception, concept development, subsequent research and collection of more specific information and an initial informal moment of viewing and tests (Reid and De Brentani 2004).

Peter Koen and his collegues (Koen et al. 1996) seeking an interpretative model, try to identify a series of differences between the FFE and NPD, aligning the phases and defining the contents, but with poor results: the two processes are characterized by a completely different nature, so the categories used to describe and assess the product development phase are in no way similar to the fuzzy nature of the first phase of the design process (Table 2.1).[8]

Koen et al. (2002) later tried to describe a non-sequential model using what is described as Bull's Eye. He identifies five FFE activities (Fig. 2.1):

1. the identification of opportunity;
2. the analysis of opportunities;
3. the generation of ideas and their enhancement;
4. the selection of ideas;
5. the generation of the concept.

[8] Peter Koen, teacher at the Steven's Howe School of Technology has written numerous essays on the FFE. One of the most well known and fundamental for the disciplinary area, written with numerous collaborators, is Koen et al. (1996).

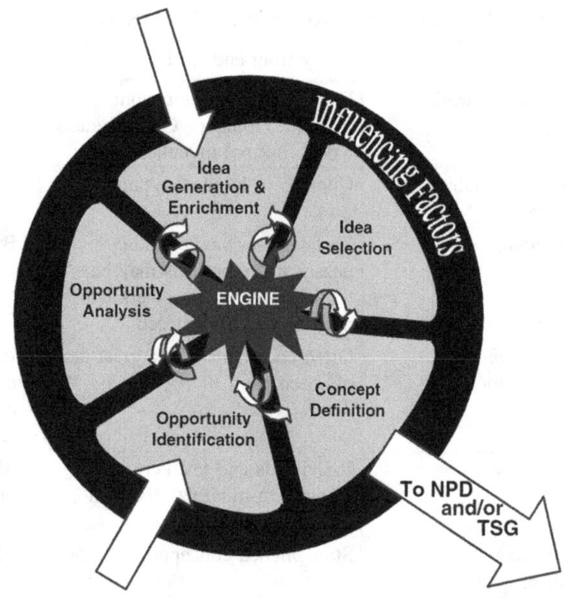

Fig. 2.1 Model for new concept development or Bull's Eye considered as relationship model (Koen et al. 2002)

These steps are placed into relation using a central engine, made up of the leadership, culture and strategies of an organization. The five steps are embedded in a series of *influencing factors,* which indiscriminately include distribution channels, users, laws, competitors, government policies, economic and political climate and the so-called *enabling sciences and technologies*.

Compared to the linear models, the Bull's Eye presents advantages: the circular shape and dual direction of relationships suggest that the process can be launched from any point, the ideas can circulate and the concepts can be reiterated. It is also possible to use certain elements of the FFE more than once and the order of the phases need not be sequential. In identifying the methods, tools and techniques that are effective through this mode, the influencing factors, while being presented as a shapeless magma in which the process is embedded, play a fundamental role. The capacity to recognize the potentials and catalyze them at just the right time is also crucial. Who plays this role of reading, interpreting and encoding the messages coming from the outside?

It is rather odd that, while placing the concept and design activities in the centre of the FFE phenomenon, the presence of design is not explained in any way. The authors of the *operation* area recognize the centrality of creativity, the experimental dimension and the capacity of single individuals to trigger the innovation process, but do not assign any role to the design discipline.

It is also necessary to observe that all recent and very recent literature on the FFE is focused on the "temporal" phase of the process. The Front End is seen, read and interpreted as a moment in its own right, which does not present particular relations with the phases and players involved in product development (SNP or NPD).

An innovation, even when it is discontinuous, cannot be kept separate from the production or distribution system, for example. Furthermore, as broadly explained in literature, innovations can be triggered at any time of the development process and consequently it is necessary to seek a critical reading key.

2.3 *AdvanceDesign* as Front End of Innovation: Object, Process, Sharing

Elisabeth Sanders, designer and researcher, founder of Maketools, has analyzed the code of design research methodologies (particularly studying ethnographic approaches and co-design), and has also sought a link between the latter FEI (Sanders and Stappers 2008). According to her vision, the FFEI, also known as 'pre-design', should circumscribe and describe the many activities that are used to instruct and inspire the exploration of open questions on future scenarios: 'how to improve the quality of life for specific categories of users (the sick, elderly, etc.), or how to identify the desires of families for their free time'. In literature, reference is often made to the FE as *fuzzy* due to its chaotic and ambiguous nature, which characterizes the initial phases of the design process.

In the FFEI, the form of the design process output is still often unknown: will it be a product, a service, an interface? Without the need to identify specific phases of the FEI, Sanders highlights that, in this crucial phase, considerations of different kinds come together: understanding the user and the context of use, exploring and selecting technological opportunities, and new materials. According to literature, the aim of these explorations into the FEI is to determine the *know what*, knowing what to design. The FEI precedes the traditional design process phases. After identifying the ideas for the product, concepts and prototypes are developed, tested and finished off on the basis of the feedback of future users.

The focus on the exploratory and non-encoded phase of the project, the capacity to positively influence the capacity for overall innovation, the role of mediation between influencing factors, research and the development of new products are strong relational elements between *AdvanceDesign* and FEI. The proven influence of FEI on the success of new product development processes has led many companies to implement an Advance Desgin centre. Well known car manufacturers, as well as Philips, Whirlpool, Samsung and Nokia, who have historically created important design centres not necessarily linked exclusively to the company itself, have acted to maximize their innovative potential. The investment in the more exploratory and not immediately strategic dimension has allowed these companies to shift the attention from design as applied research to design as an activity aimed at the future. These activities to explore Advance Desgin can be synthetically recognized as actions involving three different levels: object, process and sharing (Fig. 2.2).

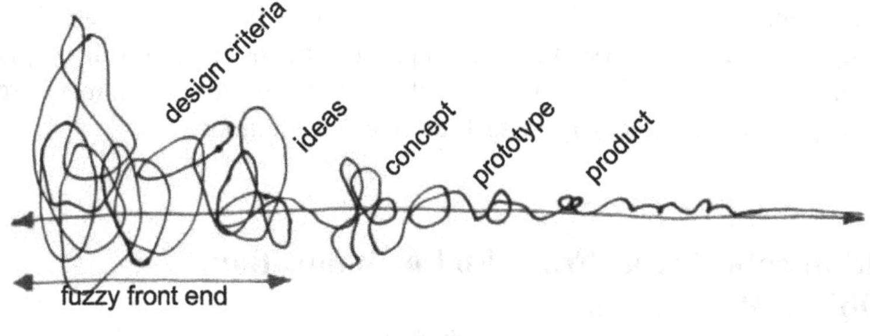

Fig. 2.2 Design and FFEI (Sanders and Stappers 2008)

2.3.1 Object: Products Versus Proposals

Plans are made for the future experiences of people, communities and cultures that are now connected and informed in a way which was unthinkable as recently as just 10 years ago. In the world of design, new disciplinary areas have emerged: the traditional areas of design (*communication-visual design, interior design, product design*) at a certain point, became insufficient to define the context of design. Just think of *interaction design,* introduced by Bill Morridge in the late 1980s, of the interest in service design of which Ezio Manzini was the forerunner, or transformation design which, already having been identified by Joseph Pine and James Gilmore, was extensively described in the white paper of the same name issued by the Design Council in 2006.[9] The emerging of these contexts indicates that design activities are not aimed only at product categories, but are extended to design by 'proposals' or 'ends'.

As emerges from the diagram shown, Sanders and Stappers highlight how reading design fields in a traditional key outlines areas, which are defined not by the outcome of the project but by the social need that generates it (Table 2.2).

Not only are the emerging areas changing what we design, they are changing how we design it, why and who for. The subject of design is undoubtedly the first visible element of this change. Historically, design is linked to tangible culture and is a tangible discipline which has to do with the material and technical dimension of the product. In its existing connotation, this same discipline, thanks to its capacity to read and interpret the evolution of the social context, acts on the intangible aspects of everyday life (services, emotions, experiences, etc.).

The first experiences of *AdvanceDesign* could undoubtedly be traced back to design activities aimed at objects. Amazing cars in which the various technologies available were brought together and allowed innovation to take tangible shape and

[9] Comparison is made between: Morridge (2007) and Burns et al. (2006) (www.designcouncil.info/mt/RED/transformationdesign/TransformationDesignFinalDraft.pdf).

Table 2.2 Comparison of traditional and emerging design practices (Sanders and Stappers 2008)

The traditional design disciplines focus on the designing of "products" …	… while the emerging design disciplines focus on designing for a purpose
Visual communication design	Design for experiencing
Interior space design	Design for emotion
Product design	Design for interacting
Information design	Design for sustainability
Architecture	Design for serving
Planning	Design for transforming

be visible to future users too. *Concept cars*, like concept products presented during fairs and events, are the visualization of future possibilities. In *AdvanceDesign* experiences, it is often as though the dichotomy between product and proposal is resolved: products projected into a remote future are designed, products that are, to all intents and purposes, proposals, crystallizing the intention of the action. Washing machines that clean using enzymes, electrical appliances that recycle water and solar powered cars are examples of tangible products, but conceived and designed for scenarios where the proposal of sustainability is central.

AdvanceDesign often places products at the centre, yet does not concentrate exclusively on the tangible dimension, but on what comes before and after the design process: it places in relation research and the production world.

The panoramas of design and design research will keep on changing due to the fact that the margins between the design activities and research activities are fuzzy. The FFE can therefore be considered the problematic sphere in which new research opportunities are outlined both for designers and for researchers: probably in a not too distant future, FEI will be populated by hybrid figures, 'design researchers and research designers' (Sanders and Stappers 2008).

2.3.2 Process: Design Process Versus Process Design

From Alexander to *operations* with their rigid method, to Bruno Munari with his culinary metaphor, to Tomás Maldonado who asserts that 'Design consists in coordinating, integrating and articulating all those factors which, in one way or another, participate in the process of formation of the object', the project has always been intended, interpreted and narrated as a process.

Today, the process is even more pervasive compared to the theme of design: if the object of design is increasingly an end and not a finished product, if objects become dematerialized to take on shapes that can be moulded in time, then design will have to take on a colloquial form, not only with users, but also with the temporal dimension. Bruce Sterling in his visionary *Shaping things* (2005) highlighted this concept, outlining a conversational design and interaction. Coining the neologism SPIME, which comes from the contraction of Space+Time, the author identifies a new type of object and proposes an approach to design which goes beyond the design of the single artefact, towards the design of a transversal experience to the different contexts of use. The subject of design, according to Sterling (2004), shifts from the artefact to the process.

> The most important thing to know about Spimes is that they are precisely located in space and time. They have histories. They are recorded, tracked, inventoried and always associated with a story. Spimes have identities, they are protagonists of a documented process. They are searchable, like Google. You can think of Spimes as being auto-Googling objects.

These are objects which merge time and space, traceable at all times, thanks to the identity and evolution of GPRS and RFID technologies, enabling a conversational relationship with people. Dialogic objects, made of material and information, the circulation of which continuously produces knowledge through what Sterling calls 'metastories'. This complex transformation, this vision of the future, described beautifully by a science-fiction author, but absolutely plausible, places man in the centre of design: no longer the passive addressee, but an active player in the process of construction of knowledge and intermediary in the creation of new objects.

As described a few years ago by Bertola (2001):

> The process of knowledge, as described by several parties, is configured as an authentic 'creative' process as, once again, it has nothing to do with knowledge of something that already exists, as considered generally, but the creation of new knowledge, thanks to the generative interaction of several individuals and players that place their tacit and explicit basin of knowledge on the web and give life to new models of reality.

2.3.3 Sharing: User-Centre Design Versus Co-design

Especially in the European context but also in the United States, design practices have been influenced by the *human centred* or *user-centred design* approach which, having begun in the 1970s and spread extensively in the 1990s, is currently considered to be one of the most effective approaches to the design and development of new products.

In the preliminary document issued by the European Union during the public consultation on design policies, it is highlighted several times how, while the idea of design is commonly associated with product aesthetics and appearance, the applications of design are much broader. User needs, hopes and desires, but also skills and abilities represent the starting point and the focus of the design activities, the capacity to combine and materialize in products, services and systems, environmental, safety and accessibility aspects without neglecting the economic components, makes design worthy of public attention.

Notoriously, certain European countries like Finland, Great Britain and Denmark have used *user-driven* or *user-centred innovation* among the mainstays of national innovation strategies. These approaches are considered as tools to make innovative products, services and systems that better respond to user needs and are therefore more competitive. In some excerpts from the document we read:

> Design as a driver and enabler of innovation complements more traditional innovation activities such as research. In the current economic climate, where resources for innovation are scarce, design and other non-technological innovation drivers, such as organizational development, employee involvement and branding, become particularly relevant. They often are less capital intensive and have shorter payback periods than, for example technological research, but still have the potential to drive competitiveness. [...] It is particularly important to find new ways to promote innovation in SMEs in low-tech sectors and regions dominated by low-tech industry—where an in-house R&D department may seem too big an investment—as well as in private and public services. [...] Design has the potential to become an integral part of European innovation policy, a building block of a policy model that encourages innovation driven by societal and user needs, and that builds on existing European strengths such as our heritage, creativity and diversity to make Europe more innovative. The development of tools and support mechanisms for design-driven, user-centred innovation, networking and research, and collaboration in education and training are areas of action that could help remove some of the barriers to better use of design in Europe (EU Commission 2009).

For some time, in practice and in design research, the user has played a fundamental role, but the theme can be looked at in two very different ways. In the reading, we can recognize two *mindsets*: on one hand, there are those who believe that the research and design activities can be carried out by 'experts' for whom users are classed as addressees, users and consumers; on the other, there is an emerging culture characterized by the dimension of participation in design, where the addressees of the design become players and co-authors in the design process.

A reading key to clarify the panorama of approaches is offered by the mapping of Sanders (2008). The map in Fig. 2.3 crosses two tensions *expert mindset*, *participatory mindset* and *design-lead*, *research-lead* tensions, classifying a series of design methodologies and tools, and especially of design research and identifying the different areas of overlapping.

The area that identifies *user-centred design* is the broadest and has roots in social sciences and behavioural sciences, in engineering, ergonomics, cognitive psychology and applied ethnography. In the majority of cases, those who operate and teach in this area have trained as researchers, not as designers.

The area that represents *participatory design* covers the right side of the map. Participatory design is a combination of theories, practices and studies aimed at involving end users in design activity. Created in the 1960s in the Scandinavian countries, thanks to cooperation with social parties, this approach has earned its own independence within Human Centred Design (HCD)[10] and, having been

[10] According to Martin Maguire the principle of HCD implicate the active involvement of users starting from the definition of the user requirements, an appropriate management and assignment of work between the user and the system, an interactive approach to design, a multidisciplinary development team. Cf. Maguire (2001) (www.idealibrary.com).

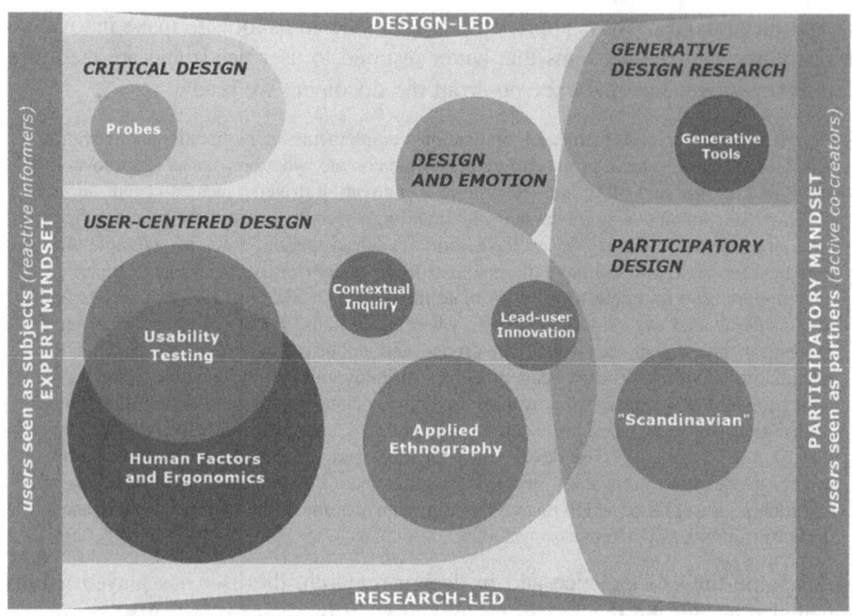

Fig. 2.3 Map in evolution for researches and practices of design (Sanders 2008)

initially spread mainly in the area of software and web design, now pervades different areas of design.

Over time, some of these methodologies have been strengthened, others have extended their range of action, but above all, the emerging design activities have concentrated in the two upper quadrants: from the scientific approach to *design-led research* or even *design-driven, user-centred innovation* as defined in the European document.

The 'Cultural probes' technique, for example transforms informers into co-designers.[11]

The exploratory experience allows the sharing of spaces and times of the investigation in the field and to reconstruct the plot of action and relations between the subjects that interact in that design context. The methodology, used in the initial phases of a design is used to generate design solutions that best suit user requirements. These techniques have recently been progressively changed and are now called 'Design probes', as in the current Philips project, oriented towards the remote future of 2020. Ambitious projects are experimenting with clothes, jewellery and intelligent tattoos, capable of adapting their design to the suit the user's mood, and they seek the consent and stimuli of the users to discover trends capable of turning into new areas of design and significant business.

[11] An investigation tool which appeared in 1999 in a pioneering study by Bill Gaver, Antony Dunne and Elena Pacenti, for the collection of experience-related material on playful, emotional and irrational aspects of the everyday life of the people studied. Cf. Gaver et al. (1999).

Equally pioneering are the 'Design documentaries' of Bas Raijmakers,[12] the 'Playful triggers' of Daria Loi,[13] the 'Mobile diaries' of Digital Eskimo,[14] the so-called 'Design games' or 'Design maketools' and the Design fiction with its diegetic prototypes (Kirby 2010). All these research tools have a common feature; they are not methods to design solutions, but methods to inspire, to generate metaphors and therefore to identify new possible roads.

References

A Study of the Design Process: Eleven lessons: managing design in eleven global brands. www.designcouncil.org.uk/en/About-Design/managingdesign/Eleven-lessons

Alexander, C.: Notes on the Synthesis of the Form. Harvard University Press, Cambridge (1964)

Asimov, M.: Introduction to Design. Pertinence Hall, Englewood Cliffs (1962)

Bertola, P.: Creatività e progetto. In: Penati, A. (curated by), *Giovane è il design*. POLI.design, Milano (2001)

Burns, C., Cottam, H., Vanstone, C., Winhall, J.: Red Paper 02. Transformation Design. RED-Design Council, London (2006)

Cross, N.: Forty years of design research. Des. Stud. **28**(1), pp. 1–4 (2007)

Design Council: Driving Recovery with Design. (www.designcouncil.org.uk/Documents/About%20design/Facts%20and%20figures/DesignCouncilBriefing_04_DrivingRecoveryWithDesign.pdf) (2009)

European Commission Working Document: Design as a Driver of User-Centred Innovation, Brussels 07-04-2009, SEC (2009) 501 final. www.ec.europa.eu/enterprise/newsroom/cf/itemlongdetail.cfm?item_id=3054

Findeli, A.: Some tentative epistemological and methodological guidelines for design research. In: Pizzocaro, S. (curated by) Design Plus Research Proceedings, Milano (2000)

Frayling, C.: Research into art and design. R. Coll. Res. Pap. **I**(1), 1–5 (1993/1994)

Gaver, B., Dunne, A., Pacenti, E.: Cultural probes. Interactions **6**(1), 21–29 (1999)

Gregory, S.A. (ed.): The Design Method. Butterworth Press, London (1966)

Hagen, P., Robertson, T., Gravina, D.: Engaging stakeholders: mobile diaries for social design. In DUX '07: Proceedings of the 2007 Conference on Designing for User eXperiences. ACM, New York (2007)

Johnson, V.S., Rumbaugh, B.A.: Educating for engineering design today measuring for excellence tomorrow. The NASA/ASRA University advanced design program. Int. J. Eng. Edu. **14**(1), 67–76, 199 (1998)

Jones, J.C., Thornley, D.G. (eds.): Conference on Design Methods. Pergamon Press, Oxford (1963)

[12] Design documentaries, according to Bas Raijmakers, are a new, visual method to discover what matters to people. They inform and inspire design processes at early stages. The method emerged from his exploratory filmmaking practice, influenced by documentary film ideas and techniques. (www.designdocumentaries.com).

[13] Daria Loi, a graduate of Politecnico di Milano, continuing her studies at RMIT where she is experimenting collaborative techniques based on 'works' that involve the user. Cf. Loi (2005).

[14] 'Mobile diaries' represent a sort of social network applied to the specific sphere of design. The creation of a network—Digital Eskimo—with shared intents (social, ethic, environmental) allowed the authors to experiment and document this method. Cf. Hagen et al. (2007).

Kirby, D.: The Future is Now: Diegetic Prototypes and the Role of Popular Films in Generating Real-world Technological Development. Social Studies of Science 40, 41–70 (2010)

Koen, P.A., Ajamian, G.M., Boyce, S., Clamen, A., Fisher, E., Fountoulakis, S., Johnson, A., et al.: Fuzzy Front End: Effective Methods, Tools, and Techniques. Industrial Research, 5–35, (1996). Retrieved from http://www.stevens.edu/cce/NEW/PDFs/FuzzyFrontEnd_Old.pdfNEW/PDFs/FuzzyFrontEnd_Old.pdf

Koen, P.A., et al.: Fuzzy-Front End: Effective Methods, Tools and Techniques. In: Belliveau, P., Griffen, A., Sorermeyer, S. (eds.), PDMA Toolbook for New Product Development, pp. 2–35. Wiley, New York (2002)

Loi, D.: Lavoretti per bimbi: Playful Triggers as keys to foster collaborative practices and workspaces where people learn, wonder and play. PhD thesis, RMIT University, Melbourne, Australia (2005)

Maguire, M.: Methods to support human-centred design. Int. J. Hum. Comput. Stud. **55**(4), 587–634 (2001)

Morridge, B.: Designing Interaction. The MIT Press, Cambridge (2007)

Nokia Press Services. www.pressbulletinboard.nokia.com/2008/04/29/homegrown—new-design-thinking-on-sustainability. Accessed 29 April 2008

Picchi, F.: Grande dizionario Inglese Italiano – Italiano Inglese. Hoepli, Milano (1999)

Reid, S.E., De Brentani, U.: The fuzzy front end of new product development for discontinuous innovations: a theoretical model. J. Prod. Innov. Manag. **21**(3), 170–184 (2004)

Sanders, E., Stappers, P.J.: Co-creation and the new landscapes of design. CoDesign **4**(1), 5–18 (2008)

Sanders, E.: An Evolving Map for Design Practice and Design Research. Interactions **15**(6), 13–17 (2008)

Shön, D.A.: Il professionista riflessivo. Dedalo, Bari (1993) [ed. or. The Reflective Practitioner. Basic Book, New York (1983)]

Simon, H.A.: The Science of the Artificial. The MIT Press, Cambridge (1969)

Smith, P.G., Reinertsen, D.G.: Developing products in half the time. New York: Van Nostrand Reinhold (1991)

Sterling, B.: When Blobjects Rule the Earth. SIGGRAPH, Los Angeles. www.viridiandesign.org/notes/401-450/00422_the_spime.html (2004)

Sterling, B.: Shaping Things. Mediawork Pamphplet Series. (2005)

US National Design Policy Initiative: Redesigning America's future: ten design policy proposals for the United States of America's economic competitiveness and democratic governance. www.designpolicy.org (2009)

Zimmerman, J., Forlizzi, J., Evenson, S.: Research through design as a method for interaction design research in HCI. In: Proceedings of the Conference on Human Factors in Computing Systems. ACM Press, New York, pp. 493–502 (2007)

Chapter 3
Maps and Tools for *AdvanceDesign*

Alessandro Deserti

> *The architect is condemned, by the nature of his work, to be perhaps the last and only humanist figure of contemporary society: forced to think the whole to the extent that he's also a sector technician, specialized, aimed at specific operations and not at metaphysical statements.*
>
> Umberto Eco

3.1 *AdvanceDesign* Between Theory and Practice

Design process has been described as a complex activity by many authors, mainly in connection with the theme of its systematic nature.[1] Such complexity or systematic nature is rendered somewhat banal when an attempt is made at reducing the process to the level of just any scheme, which does, however, help to offer a view, that can be transmitted and replicated. A model of description represents a simplification of reality, which can be assumed as a characteristic as opposed to a limit, as long as its objective is purposely didactic.

[1] The question of complexity lies at the centre of the work of different authors: Wolfgang Jonas sustains that, while the term complexity sounds promising as a departure point for the construction of a theory of design, in actual fact it is very ambiguous, and prefers to refer to the concept of 'systemicity'. This leads him to theorise that design has to develop within a system of 'gaps' in knowledge, which regard as much the present as projections towards the future, and this suggests operating through the development of numerous alternatives. On this subject cf. Jonas (2005). This conclusion, while the connection is not explained by Jonas, is very close to that proposed by 'scenario thinking', which takes the idea of the unpredictability of the future as its starting point, described in terms of likely alternatives rather than probability or determinist models. For a vision of the theme of the systemic approach (Penati 1999).

A. Deserti (✉)
Dipartimento di Design, Politecnico di Milano, Via Durando 38A, 20158 Milano, Italy
e-mail: alessandro.deserti@polimi.it

M. Celi (ed.), *Advanced Design Cultures*, DOI 10.1007/978-3-319-08602-6_3

In relation to this, it is appropriate to point out that, despite the fact that we are going to refer to different models of process and schematic portrayals, we do not intend to sustain a methodological approach; design, which has progressed through different methodological approaches[2], and which is still interpreted by many authors in this sense, can be catalogued in terms of method only within a technical view which, as we will see, occupies a position far away from the world we will be exploring in this publication.

Design, as a consequence of its open and expansive nature, is hard to catalogue in terms of disciplinary boundaries too. Even the term 'design', as quite rightly highlighted by Heskett (2002), is ambiguous: 'Not surprisingly, in the absence of widespread agreement on its significance and value, much confusion surrounds the design practice'. Also according to Heskett, design occupies an extensive space that oscillates between popular and scientific culture: while the understanding of certain disciplines does not necessarily require a background of scientific knowledge, and this makes them accessible—at least up to a certain level—also to a non-expert public; others are usually accessible only to a public with in-depth knowledge, or they require extensive introductions in order to be explained. Design is positioned ambiguously between these two dimensions:

> Design sits uncomfortably between these two extremes. As a word it is common enough, but it is full of incongruities, has innumerable manifestations, and lacks boundaries that give clarity and definition. As a practice, design generates vast quantities of material, much of it ephemeral, only a small proportion of which has enduring quality (Heskett 2002).

Design is, in fact, multiple (Bertola and Manzini 2004) and hard to outline: its boundaries are constantly changing. Looking at *advance design* means building an overall view in which to identify, by similarities or differences, an area of action in which to move. To orient these movements, we are going to use a map that creates tension between different ways of considering design (Fig. 3.1). And we are going to do it, not with the idea of distinguishing the things that are right from those that are wrong, but of understanding the meaning and the positioning, where often they cohabit in the same company or designer, of different phases of the same design process, which require different approaches and attitudes.

---

[2] The methodological approach to design, and particularly the 'first wave' of *design methods* developed in the 1960s under the strong influence of cybernetic thought, is definitely to be considered as a parenthesis, which has been overcome, of which, however, we inherit certain foundational assumptions, such as the interdisciplinary approach and the principle of *feedback*, which brings with it the relationship with the theory of systems and the idea that the output of a system, meaning its action towards the environment, falls together with other inputs in the form of feedback, allowing the system to correct its action. Born in relation to the so-called *operational research*, the approach of *design methods* found its foundation in the idea according to which mathematical models could be not only be an aid to *decision-making*, but also, to a certain extent, the theoretic foundation of design-related disciplines. This thought contrasted with the progressive inability to measure up against the variety, variability and intangibility of the needs it proposed to satisfy.

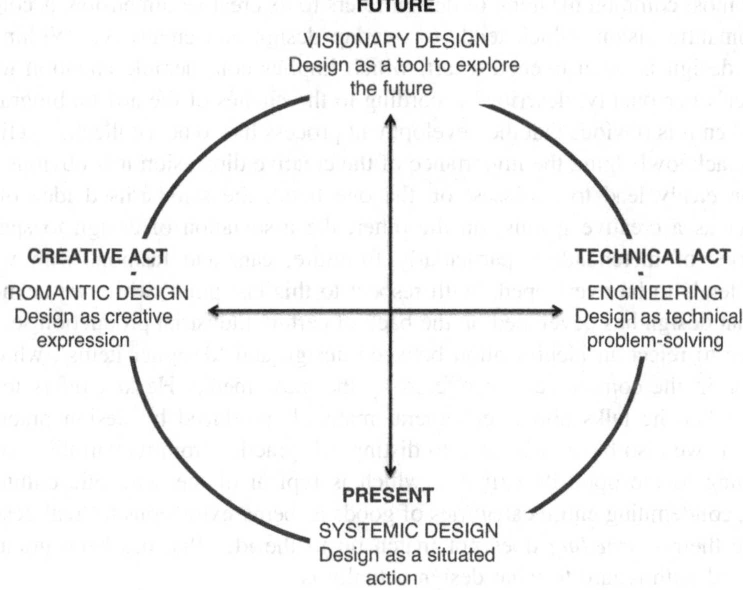

Fig. 3.1 Map of 'ways of considering design'

The map identifies two main tensions: on the one hand, that between the technical and the creative dimension, which has been extensively described as the specificity of design, to which many disciplines look to find the key to their transformation[3]; on the other, that between the observation of the present and the exploration of the future, which reflects the tension between the urge for answers to contingent needs and that of moving away from the everyday situation.

[3] The theme of *design thinking*, considered as a transferrable approach from the world of design to other disciplinary environments, mainly those that refer to economic science, is based upon the consideration of the fact that design is positioned midway between technique and creativity, from which it has developed a qualitative and holistic approach, which is particularly interesting insofar as determinist approaches have difficulty in synchronising into the variety and variability of the real world. Moreover, design appears to be historically 'trained' to deal with uncertainty, which seems to have become the permanent condition in which companies find themselves having to operate and in which managers have to take decisions: 'Today's markets are increasingly unstable and unpredictable. Managers can never know precisely what they are trying to achieve or how best to achieve it. They cannot even define the problem, much less engineer a solution. For guidance, they can look to the managers of product design, a function that has always been fraught with uncertainty' (Lester et al. 1998, pp. 87–96). And also: 'The style of thinking in traditional firms is largely inductive—proving that something actually operates—and deductive—proving that something must be. Design shops add abductive reasoning to the fray—which involves suggesting that something may be, and reaching out to explore it. Designers may not be able to prove that something is or must be, but they nevertheless reason that it may be, and this style of thinking is critical to the creative process' (Martin 2009).

The most common meaning of design refers to its creative dimension, according to a romantic vision, which tends to overlap design and creativity.[4] Within this vision, design is often likened to art, which implies considerable attention to the designer's personality, described according to the clichés of the artistic biography, even when it is obvious that the development process has to be a collective activity. Despite acknowledging the importance of the creative dimension it is obvious how this can easily lead to excesses: on the one hand, the standardised idea of the designer as a creative genius; on the other, the association of design to specific categories of merchandise, particularly furniture, cars and fashion, with which design tends to be overlapped. With respect to this last point, while we acknowledge that design has developed on the back of certain industrial production sectors, we have to reject an identification between design and 'designer items', which is common in the communication offered by the mass media: Heskett refers to this (2002) when he talks about 'ephemeral material' produced by design practices. However, we also have to be able to distinguish practice from its narration, or we risk going to the opposite extreme, which is typical of the academic culture of design, condemning entire categories of goods as being extraneous to 'real design', because their *storytelling* does not match up to the idea that has been gradually developed with regard to what design actually is.

Contrary to the romantic approach, the engineering approach may be categorised within a positivist view, according to which design can be described as a rational sequence of steps that lead from the formal definition of a group of requirements to the development of a solution, which gains in validity in relation to its ability to satisfy the preset requirements. Looked at like this, design is often treated in methodological terms, paying considerable attention to the rationalisation of the process, with the idea that the method is necessary to the optimisation of the designer's performance.

Another meaning, described by Fallman (2003) as 'pragmatic', highlights the 'situational' nature of design. According to this view, design does not follow preset models, but tends to adapt to situations, altering paths and sets of tools to suit the context in which it has to operate. According to Fallman, the 'pragmatic' approach to design is configured as a hermeneutic process of interpretation and creation of meaning:

> Rather than science or art, […] design takes the form of a hermeneutic process of interpretation and creation of meaning, where designers iteratively interpret the effects of their designs on the situation at hand.

[4] The 'romantic' approach to design is described by Daniel Fallman, who offsets against the 'rationa' approach, similar to what we defined as the 'engineering' approach and the 'pragmatic' approach, which has a hermeneutic dimension linked to the interpretation and creation of meaning. On this matter, cf. Fallman (2003, pp. 225–232).

In our view, the innovation of meaning, which according to Klaus Krippendorff is the real objective of design,[5] represents just one of the possible paths of innovation by design: possibly the most complex, but definitely not the only one possible. If we look closer, 'situational' design, being set in a tangible context made up of places, people and firms, is closely linked to contingency and pays little attention to exploring the future. In this sense, we find it hard for it to identify with the capacity of building new meanings.

Design can also be seen as a visionary activity, where particular emphasis is placed on the ability to explore the world of what is 'possible'. This way of considering design is usually strongly linked to the time factor: the shift in the time axis on which we are required to operate in terms of design tends to significantly condition the designer's attitude and the meaning of design itself. Exploring the future is the job more typically associated with *advance design*: the further the time axis shifts, the more it is necessary to bring specific visionary capacities into play and the more we move away from contingent problems to the investigation of possible worlds. In actual fact, as we have been able to highlight, exploring the future is just one of the ways of intending *advance design* which, more specifically, has to do with the ability to develop new visions, or to generate a shift, which does not necessarily have to be projected in time.

If we go back to our map, *advance design* falls within an area which cannot be perfectly outlined, and takes on different characteristics depending on its proximity to the different approaches to design that we have described, which act as 'attractive poles' around which it gravitates. Of course, if we look at the axis that links the exploration of the present to that of the future, *advance design* tends to be more attracted by the future than the present. In the same way, it tends to move within the area of creativity more than that of a technical-engineering approach. However, a vision completely focused on design is inadequate: if we look at the relationship between design and other disciplinary spheres, *advance design* preferably frequents border areas rather than the centre of the discipline, with all the opportunities and dangers that this position implicates.

An almost 'topographic' vision of the relationship between disciplines helps us clarify the point. It is as though the disciplinary territories could be represented by concentric centres: in the middle, we find their tradition and as we move towards the outside, we find new trajectories and new environments. The path from the centre of the circle to its circumference implicates a progressive reduction of the possibility to make reference to the consolidated disciplinary knowledge, a shift of the axis of interest and operationality towards new environments and the need to

[5] Entering the literature that considers the design disciplines in relation to semiotics, Krippendorff sustains that the construction of revealing of meaning has always been the main purpose of design: 'Looking back, Industrial design has always been concerned with what industrial artefacts mean. All schools, all movements, all philosophies, however short-lived or ill-conceived they may appear to us now, can be characterized by their particular approach to making sense of material culture' (Krippendorff 1995, pp. 138–162).

look at what other disciplines are doing, dealing with similar problems, trying to transfer models of interpretation and action.

In this topographic vision, we can identify more static disciplines, solidly anchored to their tradition, and expansive disciplines, which tend to progressively expand the dimension of their circumference. Observing what happens from above, we can see that, starting from the closest positions, some circles tend to touch others, intersecting with one another, generating overlapping areas. Obviously these are places frequented simultaneously by several disciplines: new fertile grounds, virgin spaces to be conquered beyond the frontier; places of exchange, where you can understand how it is possible to work together to plough the ground and harvest the fruits; places for confrontation, where there is no intention to cohabit, sharing the available space.

The expansive, static or recessive nature of the disciplines seems partly innate in the way they are (some are born closed, others open, while others alternate open and closed moments) and partly linked to contingencies. Design, which is born as an open discipline, is currently going through an undoubtedly expansive period: on the one hand, the number of sectors that look to design as a fundamental component for building competitive capacity, attributing meanings and satisfying needs is increasing; and on the other, design moves proactively, seeking new areas of operation, exploring border territories and assigning itself new roles.

The most significant danger in the frequenting of these outlying places is that of losing contact with the centre, making your identity fade and finding yourself speaking the language of the others.

Design now frequents these territories of encounter and confrontation on a daily basis, experiencing constant incursions and more or less peaceful invasions: it works on strategies and invades the territory of marketing; it analyses users and invades the territory of *humanities*; it takes care of art direction and invades the territory of communication; it develops technical solutions and invades the territory of engineering…But perhaps the most worrying thing is the inverse phenomenon: while, when listening to someone who declares himself to be an expert on physics, engineering, medicine or sociology, we are reasonably certain that we are listening to a physicist, and engineer, a doctor or a sociologist, when we listen to someone talking about design, especially in theoretic terms, we can be reasonably certain that we are rarely in front of a designer. This is both an opportunity and a problem.

Many design theories make reference to extra-disciplinary know-how, or explain design in relational terms, relating to other disciplines.

The numerous attempts to develop interpretations or determinist theories of design, in which it is read in relation to a 'single' or prevalent factor (form, function, value, sense, etc.), reveal how it can be contaminated by proximity to one or another disciplinary sphere, but they definitely do not work when they want to become 'absolute' models.

Multidimensional or multifactorial models are probably closer to the systemic dimension in which design is naturally immersed, but it is hard to translate them into tools that can be used by those who operate in the design field.

Even user studies, regardless of the large variety of possible approaches, which span from *humanities* to marketing and ergonomics, represent one of the possible accents of design, but they definitely do not exclude the possibility of tackling it from other complementary or alternative points of view.

In this 'relativist' context, the temptation to generalise takes shape, finding sufficiently broad models of description to comprehend everything that design is today and will be tomorrow, which solves certain problems while generating others. For example, we have already mentioned *design thinking*, a term on the use of which there is now a certain consensus, to the point where different theories have grown up around it, usually outside the world of design: while it is reasonable to assume that design can 'also' be described as way of thinking, it is not quite so reasonable to sustain that design is a way of thinking and that anybody can be a designer.

In the same way, the identification between the theory of design and that of innovation does not work: on the one hand, it is obvious that design has deep relationships with innovation, but that it cannot completely overlap with it; on the other, this type of approach leads to qualification of the contribution of design in relation to the types of innovation. The association between design and innovation often generates models of interpretation that tend to overestimate the individual dimension—whether it is that of the designer or that of the firm—with respect to the collective dimension, and the *breakthrough* approaches with respect to those of an incremental nature, which maybe makes sense if read from the point of view of studies on innovation, but which is not very pregnant if read from the point of view of design. And, while we are talking about the distinction between radical innovation and incremental innovation, it should be said that this topic is fed by new argumentations, but it is not particularly new in the world of design:

> No design can exist in isolation. It is always related, sometimes in a very complex way, to an entire constellation of influencing situations and attitudes. […] Earlier generations solved this problem by using many hands and minds over periods of centuries […]. The 'designer' then was not an individual, but an entire social process of trial, selection and rejection. Today, he is still that, though in a somewhat different sense, and we tend to overestimate his significance as an individual (Nelson 1957).

Nelson (1957) sustains that it is not possible to qualify design through the subject of the design process, which leads to associate innovation to the success of design. To clarify the point we can use the example of the vase, which definitely is not a new object, sustaining it is possible to design a new vase creating an outstanding example of design: the success of the design does not depend on the object, but on methods and results.

This way of reasoning is also valid particularly for the sphere discussed in this publication: *advance design* has nothing to do with particularly advanced or 'frontier' design themes, and not even with necessarily focusing on a *breakthrough* innovation. It concerns the capacity of developing innovation by looking at design objects and subjects in a new way. To do this, those involved in the world of *advance design* need to trace a path and use navigation tools to reach their destination.

Once again, we are aided by the topographic metaphor: moving away from the roads we have travelled every day for years, it becomes necessary to draw up new maps that allow us to travel within unfamiliar territories, in which it is not possible to move 'by memory' without risking getting lost. Hence, the possibility of exploring the boundaries, when the conditions are right: the availability of maps, the ability of using tools to trace the route, the ability of speaking other languages and communicate with those we meet in the boundary areas or who visit our territories. In nautical terms: we find ourselves switching from visible navigation to instrumental navigation, which is necessary at least until we become familiar with the new territories, but at this point even new territories will probably be undergoing exploration. In these terms, it seems that the expansive process is limitless and the growth of the 'design territories' is destined to keep going on forever. Obviously this is not possible: the disciplines expand and stabilise, they recede and expand again, are born and then die. Even design will probably have its pillars of Hercules. What we are interested in today is not understanding where they are, but creating the maps and tools to navigate.

3.2 Tools as a Qualifying Element

If design cannot be defined simply as a form of thought, and if its ephemeral boundaries make clear identification difficult, a significant contribution to its qualification definitely comes from the peculiarity of the skills of design operators and the tools they use.

Skills and tools are not just functional data, but elements that make a decisive contribution to defining design: It is as though we are overturning the traditional point of view, which starts from the theoretic basis and progresses through to practice, observing design from the opposite angle, which allows us to build a theory starting from practice or to concentrate on practice, forgoing the idea of building up a unique and univocal theory.

In design, the way we use tools has a combinatory nature: those who work in the design field tend to use a set of exploration, mediation, projection, narration and synthesis tools that can be rearranged depending on the context in which they are required to operate and the specific need to be met.

This set has a dynamic nature, not only in daily use, but also with reference to its progressive quantitative growth, which corresponds to the progressive expansion of the field of operation of design and the continuing shift of the frontiers mentioned before. Moreover, it would be wise to look at how the tools used by design have (at least for now) a low level of standardisation (and sometimes even formalisation), where different operators at different levels use similar but different tools.

The use of design tools is relatively free: unlike other disciplines, design tends to avoid excessively rigid patterns. Tools represent resources available and can be employed differently depending on the specific nature of the situation to be tackled. In this sense, design, more than other disciplines, is able to economise on resources, gauging the set of tools to be used in relation to the availability of resources that can be employed. As it is unnecessary to apply a standard approach and having developed the habit of adapting the set of tools to the situations, design is used to operating without big investments in research and development, leading to frequent operations with small firms. To be precise, particularly in Italy, the fertile ground in which design is cultivated is that of small and medium enterprises. In some sectors, which at a certain point became known as *design-oriented*, these firms, which have only occasionally frequented other disciplines, have developed a strong relationship with design, which has become the engine of innovation, a tool of recognisability and the mean through which emerge and stand out on the international markets.

The subject of tools also has to be categorised in the 'open' nature of design which, as we have seen, tends to escape confinement logics, which often leads to a more or less relaxed use of tools born in other disciplinary spheres, which are reinterpreted in a different culture and using different skills.

The *set of tools* used in different situations is also, to some extent, a design qualification element. The metaphor that best clarifies the concept is that of the doctor: sometimes he operates in a big organised structure (the hospital doctor); sometimes he just has a bag of instruments (the emergency doctor); sometimes he operates on a general basis (the family GP), using basic instruments; sometimes he is vertically specialised (the medical specialist), and uses specific and sophisticated instruments.

3.3 Design as Tension Between Restrictions and Opportunities

Design lives historically in the middle of two complementary and contradictory dimensions. On one hand, design as a tangible discipline, focused on the technical dimension of product development, used to dealing with the tangibility of artefacts, using materials and governing production technologies. On the other, design as a creative discipline, focused on inventing new forms, ways of use and meanings, used to dealing with the intangibility of needs and desires, studying and moulding only partially tangible features.

In a certain sense, it is as though the designer represents the link between the *back office*, where desires are focused on and take shape, the factory, where they materialise, and the points of sale, where they are made available to the market.

This intermediate position not only represents a problem, but also an opportunity, which can be taken, because of all the disciplines that become part of the product development process or—more broadly—the mechanisms of innovation, design is the only one that retains a position midway between hard and soft, while the others tend to concentrate on one of the two extremes. In a manner of speaking, we are not far from the idea expressed by the semiotics of architecture when they painted the architect as the last humanist, or, to quote Eco (1968), as an intellectual sentenced to humanism.

Many 'historical' design theories are structured as multidimensional, outlining a professional figure and behavioural models, which keep technical-economic, social and market dimensions together. With respect to this past, however, many things have changed and today's designers find themselves having to operate in a very different setting. This distance between design venues and the factory have increased: the presidia of the tangible dimension while not having disappeared has progressively become less important if related to the strategic assets: communication, distribution, brand and consumer experience.

Within this setting, design as it is traditionally understood—design that 'designs' products—is joined by countless new areas of interest, points of view, process models, tools and professional figures. Consolidated know-how is increasingly less able to guide the everyday activities of those operating in the design field, switching from concentration on the tangible dimension to operation, which extends more and more towards what comes before and after the product, areas of interest mainly governed in a more or less explicit way by other disciplines.

The tension between restrictions and opportunities is simultaneously the reflection of the technical and the creative dimensions of design and that between present and future, where we tend to look for restrictions in the present and opportunities in the future; and in the same way we try to employ the creative dimension in revealing and exploiting opportunities and the technical dimension in building a functional response to restrictions.

In the latest reading into innovation, the two souls of restrictions and opportunities tend to be described as separate: on the one hand, we have the Front End of Innovation (FEI), chaotic and distinguished by the creative dimension; and on the other, New Product Development (NPD), organised and distinguished by the rational dimension.

Design can, therefore, be represented once again as the result of a constant tension between the world of restrictions, or the limits within which every design-related action has to move, and the world of opportunities, or the trajectories of innovation which can be followed, regardless of or despite restrictions (Fig. 3.2).

The world of *advance design* is positioned within this tension, characterised differently also in relation to the prevalence of one dimension or the other. The more we imagine that certain restrictions can be, at least momentarily, set aside, the more we are able to generate and practice new scenarios, positioning ourselves in a dimension of anticipation. The more, however, we keep our feet planted firmly on the ground, anchored to the need to give a response to technological, productive and market needs, the more we find ourselves giving tangible answers to contingent needs.

It is somehow as if we could describe a sort of scale of design-related freedom, which has its lowest value when design is closer to *engineering* and its highest value when it is positioned within *advance design*.

This same model of interpretation is often applied to the design process, in which the degree freedom is progressively reduced, switching from the exploration and conceptualisation phases to those of development, engineering and industrialisation. While this is a didactically effective model, we know that a linear process based on the idea of the progressive refinement of design into a sequence of phases, does not fully grasp what really happens, since the design process becomes filled

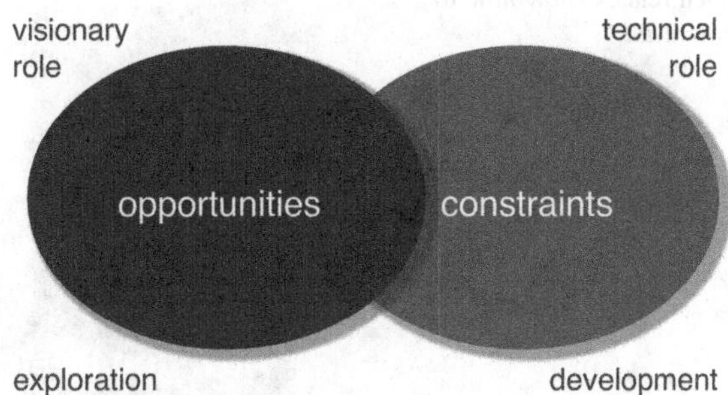

Fig. 3.2 Design tensions

with mistakes, experiences moments of crisis, often needs to develop research and conceptualisation activities continuously as opposed to in the start-up phase only.

The widespread use of the linear model is linked to a series of concomitant factors:

- the fact that literature is usually written following the analysis of successful cases, which are often those in which it is easier to reconstruct a hypothetically linear path, which leads to an effective solution, while in actual fact there are probably far more unsuccessful cases, in which the linearity of the path is much less legible[6];
- the fact that linear models are those that can be described more cheaply, starting from the assumption that they can be portrayed as a sort of funnel[7] in which, in subsequent phases of refinement and filtrations, a chaotic system of information is condensed into a product;
- the fact that the need to talk about the innovation process or, more simply, the product development process, does not fall within the scientific investigation sphere, but between this and business consulting, which leads to the attribution of an ideal character to the models rather than a portrayal of reality;
- the fact that the linear model often simplifies the narration of the process of innovation, presenting it as a simple journey, when it is often a road, which does not lead to the destination, along a path as flat as people might lead you to believe.

To understand the reasons of the non-linearity of the process, we also have to look at how often in design only the starting point is known, while the arrival point

[6] Here, it is appropriate to look at how literature usually employs unsuccessful cases as a tool to help strengthen, by difference, the idea that the design path can be traced back to its sequential and rational dimension. Just think, for example, of how the cost spiral is typically described as the result of the attempt to change design choices too far on in the development process, rather than as the result of an incorrect scenario or of conceptually inadequate solutions for the reference scenarios.

[7] The *funnel* is one of the typical portrayals of the design process, which presents the progressive refinement of design solutions, the progressive focusing of the subject of design and the progressive reduction of the informative chaos, which lead to the production of a 'distillate' representing the outcome of a synthesis process.

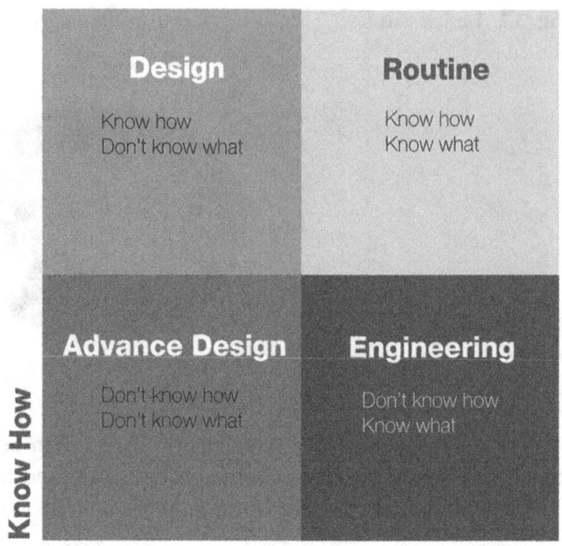

Fig. 3.3 The position of *advance design* in a matrix which relates *know how* to *know what*

is not clearly defined. This is a tautological consideration to a certain extent: if the arrival point were already known, the design would be useless. Design paths can differ greatly depending on the knowledge one has, or would like to have, of the arrival point. This too is an element that helps us categorise the subject of *advance design*: we can work on the design of a car, which has to be on the market in 24 months, and in this case we definitely are not in the world of *advance design*; or we can work on the design of a car we can imagine in the more or less distant future and, in this case, we are operating in *advance design*. Within the same possible future, we can also work on the design of new individual mobility solutions: we will not necessarily design cars in this case, but we are definitely working with *advance design*. To go back to our starting point, if we were to imagine that it is possible to define several progressive levels of *advance design*, the higher levels are definitely those in which the design subject is less defined.

If we were to position ourselves in a matrix which relates *know how* to *know what*, *advance design* definitely practices the area in which previous knowledge is not so readily available or in which it has been purposely bypassed (Fig. 3.3).

3.4 A Process Model for *AdvanceDesign*

As we have tried to explain, the trajectories of a design path can be macroscopically built, on the one hand, with the capacity to analyse and interpret the restrictions that the context in which we operate places us; on the other, through the capacity to grasp or generate opportunities, channelling creativity within the trajectories of innovation.

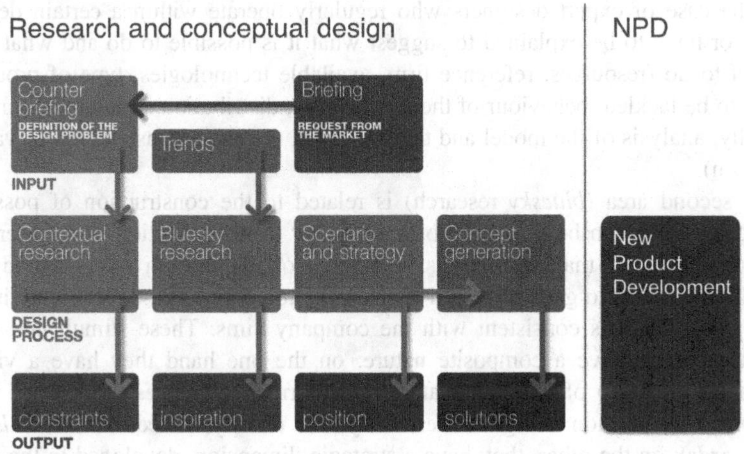

Fig. 3.4 A process model for *advance design*

As already mentioned, this is a scheme, which only partly grasps the complexity of the problem, where we ought to look at how every restriction can be potentially transformed into opportunities for a designer. The process model we are going to propose does not intend to be a rigid tool, but a tool which allows us to work around the tensions we have tried to highlight, with particular reference to that between restrictions and opportunities (Fig. 3.4).

The design process usually starts with a briefing, which is generally set up outside the design functions and is entrusted as the first element to be addressed. The briefing can be an informal and unfocused indication, or a combination of particularly rigid specifications, depending on the sector and the design attitude. Because the result of the design is strongly conditioned by its premises, the more the design intends to be advanced, the more it is necessary to offset the briefing with a 'counter briefing', stating that, on the contrary to the briefing, it is set up within the design functions, in a dialectic logic between the designer and the customer, inherited from the tradition of Italian design.

The design path on the hand is articulated, at least up to the level of conceptualisation, in four steps which mix exploration and development activities.

The exploration of the limits and opportunities is structured into two macro-areas, which allow us to make a functional simplification in order to clarify our argumentations.

The first area (contextual research) regards the understanding of how it is possible to derive information to help address the design activity from the design environment and from the players involved in the innovation process. This information defines the perimeter, or the combination of limits, which are sometimes implicit, insofar as they are already known to those who tackle the design activity, as in the case of expert designers who regularly operate within a certain design sphere, or have to be explained to suggest what it is possible to do and what it is best not to do (resources, reference firm, available technologies, type of product, market to be tackled, behaviour of the competitors, distribution structure and, more generally, analysis of the model and the players in the chain/constellation of value, and so on).

The second area (*bluesky* research) is related to the construction of possible directions, which can be practiced to innovate, or to the definition of a system of opportunities rather than restrictions. This area of exploration offers a body of helpful information to generate creative sparks or to control and direct creativity in such a way that it is consistent with the company aims. These stimuli and suggestions actually have a composite nature: on the one hand they have a visual dimension, made up of cultural, material and formal references (which are very widespread in fashion design, where they are usually called *moodboards* or *trendboards*); on the other, they have a strategic dimension, developed in the next step, through a series of 'maps of innovation', which we call scenarios, built up through the interpretation of strong and weak signals derived from observing the evolution of goods, behaviours, markets, technologies, etc. At this level we find stimuli capable of defining the strategic orientation of a firm; while at the previous level there is a repertory of suggestions that allow arrangement in a system of consistent products, which is the fundamental aim of the conceptualisation phase.

The first exploration phase is usually flanked by a pre-design or meta-design phase, which we normally call *scenario building*. The scenarios are typically a meta-design elaboration, in the form of *storytelling* of one or more possible (future) design environments, aimed at defining innovation trajectories. They do not represent a specific solution, but aim to guide the development of innovation. In the design world, scenarios are usually made up of graphic maps, which form a sort of topographic representation of innovation, making it possible to localise a departure point and trace a route towards a hypothetical arrival point. In some cases, their nature is more explicitly and intimately linked to the language of design. Sometimes, there is an overlapping between the phases of the scenario and conceptualisation of solutions, which tend to integrate until they become a single entity. This approach historically represents one of the typical territories of *advance design*. Think once again of the car sector, which expresses scenarios mainly through the development of prototypes (known in the past as *dream cars* and now usually known as *concept cars*), which anticipate a possible future with meta-design aims, or as trajectories of innovation. These products are definitely not destined to the market, but to circuits of innovation and sector-specific media, in relation to which they are presented as declarations of intent and sometimes as stores of solutions destined to subsequently enter the market, or address the development activity of the products concerned.

References

Bertola, P., Manzini, E.: Design multiverso: appunti di fenomenologia del design. POLI.design, Milano (2004)

Eco, U.: La struttura assente. Bompiani, Milano (1968)

Fallman, D.: Design-oriented human-computer interaction. In: CHI '03. Proceedings of the SIGCHI Conference on Human Factors in Computing Systems, vol. 5, n. 1, pp. 225–232, Fort Lauderdale, Florida, April 5–10, 2003

Heskett, J.: Design. A Very Short Introduction. Oxford University Press, Oxford (2002)

Jonas, W.: Designing in the real world is complex anyway—so what? Systemic and evolutionary process models in design. In: Proceedings of the ECCS 2005 Satellite Workshop "Embracing Complexity in Design", Paris, November 17, 2005

Krippendorff, K.: Redesigning design: an invitation to a responsible future. In: Tahkokallio, P., Vihma, S. (eds.) Design—Pleasure or Responsibility?, pp. 138–162. University of Art and Design, Helsinki (1995)

Lester, R.K., Piore, M.J., Malek, K.M.: Interpretive management: what general managers can learn from design. In: Harvard Business Review, vol. 76, n. 2, p 87–96, March–April 1998

Martin, R.: The Design of Business: Why Design Thinking is the Next Competitive Advantage. Harvard Business Press, New York (2009)

Nelson, G.: Good design: What is it for? In: Problems of Design. Whitney Library of Design, New York (1957)

Penati, A.: Mappe dell'innovazione—Il cambiamento tra tecnica, economia, società. Etas, Milano (1999)

Chapter 4
Understanding the Past While Planning the Future: Times and Ambitions About *AdvanceDesign*

Giulio Ceppi

> *I would like to try to demonstrate how in a certain sense time precedes the universe: that is, the universe is the result of an instability that occurred from a situation that preceded it: therefore, the universe is the result of a large-scale phase shift.*
>
> Ilya Prigogine

4.1 Creativity and Design: A Changing Scenario

1. Nowadays, it seems like creativity is being forcefully glorified, as if designers and architects were *superstars* precisely because they are portrayed by the media as being creatively gifted individuals.
   On the contrary, creativity is not a trait that can be decided by others, it is instead a job, a daily schedule, a continuous research, a new and challenging balance between several factors: as Italo Calvino wittily said, "Fantasy is like jam…You have to spread it on a solid piece of bread. If not, it remains a shapeless thing…out of which you can't make anything."
   What really matters for a designer is the path of discovery, his attitude towards the unknown, his exploratory capacity, and not so much his communication skills and the number of interviews given or published work.
2. Today, design deals with the scale of complexity and globalisation, by acquiring the relevant tools and appropriate logics, well beyond the narcissistic acclamation for personal style, for fashionable showrooms, but rather towards a more systemic and open vision, more attentive to the evolution of the subject than to a sole epiphenomenon.

G. Ceppi (✉)
Dipartimento di Design, Politecnico di Milano, Via Durando 38A, 20158 Milano, Italy
e-mail: giulio.ceppi@polimi.it

M. Celi (ed.), *Advanced Design Cultures*, DOI 10.1007/978-3-319-08602-6_4

Design goes back to dealing with foundational and unexplored matters, it does not act solely on language and trends, but on problems and hidden issues, on conflicts, and on deep and genuine emotions, it chases feelings rather than pieces of news, thoughts and not just emotions.

3. Historically, design was born to answer questions, at a given brief. Today, it wants to reformulate those questions, create new work groups.
   Design began as a specialized and individual activity. Today, it wants to be humanistic and interdisciplinary, collective and participatory. Design was created to deal with production and industry, to shape machine products. Today, it wants to restore the balance between consumption and society, giving new shape to the processes of production.
4. In the past, design dealt with functional and hands-on aspects. Today, it looks like the subject of an exhibition, of a magazine cover, more connected to the aesthetics of communication than to functionality. Design took care of needs. Today, it seems like there are only desires and they have become our new needs, but we also have real needs that perhaps have been lost and forgotten, and which we no longer perceive.
   In summary, design used to have a social purpose, useful in teaching about consumption and suggesting the new: today, all too often, it seems it just wants to amuse, amaze and entertain. We want design to go back to educating, sensitizing, and making people think and discuss, creating new relationships and developing themes and issues that are relevant and useful, on a cultural level and as models of development.

4.2 Five Arguments for a Praxeology in *AdvanceDesign*

From the above, we can conclude that the current scenario in design seems to embody what Jean Baudrillard had already anticipated back in 1987:

> Today, what we are experiencing is the absorption of all virtual modes of expression into that of advertising. All original cultural forms, all determined languages are absorbed in advertising because it has no depth, it is instantaneous and instantaneously forgotten. Triumph of superficial form, of the smallest common denominator of all signification, degree zero of meaning, triumph of entropy over all possible tropes (Baudrillard 1987).

I want to believe (and perhaps demand) that in the near future designers will not justify as much their actions in achieving a commercial and advertising objective decided beforehand, but rather in the production of a new downstream value, in the capacity to trigger a process that strengthens the identity of the interlocutors involved, whether it is the company or the consumer, public entities or private companies. In this purely ethical and social dimension, but at the same time subjective because it is driven by individual will, can be justified the meaning of the word design. So what are the value attributes that design is able to pursue and produce with its eccentric work (but coordinated) compared to the corporation or institution with which it operates?

Answering these questions is a task which is beyond this piece of writing, but we can recommend useful tools and procedures to address the issue, considering them as part of what we intend as *advance design*. We could say, by paraphrasing Zygmunt Bauman, that *advance design* is just a 'form of art' for those who deal specifically with design, to be organized and in organizing:

> We are all artists of life: whether we know it or not, will it or not and like it or not. To be an artist means to give form and shape to what otherwise would be shapeless and formless. To manipulate probabilities. To impose an order on what otherwise would be chaos: to organize an otherwise chaotic—random, haphazard and so unpredictable—collection of things and events by making certain events more likely to happen than all others (Bauman 2009).

Therefore, *advance design* is not if a technique to control probabilities, a form of organization of possibilities.

Bellow, to demonstrate this, we list five arguments of high priority and importance to try, not so much a definition, but rather a praxeology, perhaps we should say an aesthetic of *advance design*, which in itself we would want to be almost indefinable except in implementing the design.

4.2.1 AdvanceDesign Bandwidth: Set Time at the Centre of the Design

Advance design could be understood and appreciated as if it were a stratagem, a cognitive tool, a metaphor to address the design system with fresh eyes, trying to find in the profession a different and complementary role to the media spectacle, in which it becomes a format of itself. The briefness of our time is a condition of design and a limitation to what we can 'do', as rightfully explained by Richard Sennett:

> If in this way cultures' time is short, in another way it is long. Because cloth, pots, tools and machines are solid objects, we can return to them again and again in time; we can linger as we cannot in the flow of discussion. Nor does material culture follow the rhythms of biological life. Objects do not inevitably decay from within like a human body. The histories of things follow a different course, in which metamorphosis and adaptation play a stronger role across human generations (Sennett 2008).

Through *advance design* we must rethink 'from the inside' the social and economic role of design, its ability to convey information and connect knowledge. If we used the language of information technologies, we would say that the extent of design must be increased (design bandwidth), and that it is necessary to discuss the broad and complex scenarios that open before us as we move forward guided by the freedom to explore two generative axes of great strategic value (see Fig. 4.1).

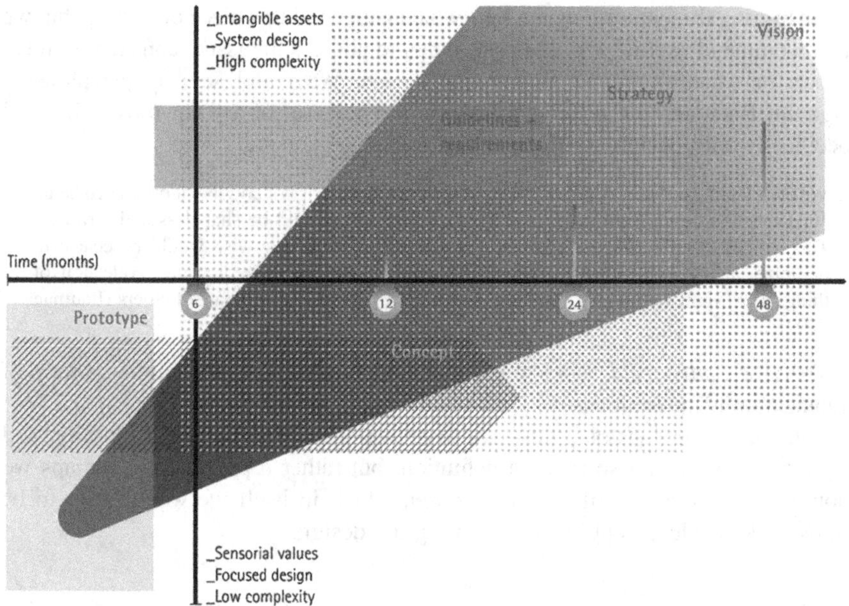

Fig. 4.1 Polar graph capable of distinguishing different types of 'design flow': in complexity and ambition (ASSETS/*vertical axis*) and in temporal spectrum and predictive capability (TIME/*horizontal axis*) (TT BANDWITH, 2003, by Total Tool, www.totaltool.it)

4.2.1.1 Time Axis

Discussing time variable in design signifies having to address a complex issue, but we can readily distinguish at least three fundamental questions:

- *Implementation expectations*. When will my design be implemented/produced? In 6, 12, 24, 48 months? A prototype has an implied or inferred time limit different from a concept, times for development and possible transferring to the market that are very specific and unique, that need to be understood and discussed beforehand, but without claiming to be absolute in the expectation.
- *Duration*. How long will my design take? How can I predict its obsolescence? A mobile phone instead of the packaging of a detergent tends to have their lifespan determined by the laws of the market: planned obsolescence, *refresh, restyling*, etc. can and must be frequently defined upstream. Obviously, the theme includes aspects related to replacement and implementation, rather than on disposal and recycling, in accordance with the principals of eco-design;
- *Management method*. How do I want to manage and transform my design as it unfolds and with which tools will I be checking its progress? A design is not a series of technical papers in plan view or section, but an open set of instructions to understand the cultural, commercial and communicative nature of an object/service.

To move along the time abscissa can also involve the synchronic consideration and the coexistence of these variables. I can develop a product concept for 24 months (implementation), imagine that the product lasts 3 years on the market with two stylistic variations (duration), therefore, know that I will need to produce for my client further phases of work to be programmed in detail in timing and content (management).

4.2.1.2 Systemic Axis

Facing the specificity of a topic in a timely and targeted manner, epiphenomenal and local, or open to a systemic and complex horizon, detailed and layered, may be necessary and correct approaches, but certainly different in ambitions and creative procedure.

Take the example of the *urban ticket* and access to city centres by using electronic payments or passive charging systems, tried by the Company Q-Free on behalf of the consulting firm Total Tool. The action plan could focus on the design of a technical component or functional detail (such as an interactive unit for automatic payments of toll fees), engaging in its own specific research on formal configurations, materials and related production technologies, rather than on the systems of interface or command, on the finishes and the product graphics, on ergonomics of use and reading of information inside the vehicle. It is about developing a specific and well defined product, although different levels of complexity may be identified.

Certain products may be designed for rapid time-to-market, others for a catalogue in the near future to be defined by ad hoc customization for various clients, yet others may be concepts to show potential future evolutions of the system, in which case by inserting new technologies, materials, ergonomics, and therefore reopening the initial brief.

The same topic can be addressed instead at a systemic level: the evolutionary scenarios of urban transport and the trends on social mobility, the modes of relationship between the political and administrative systems, the creation of a catalogue offer suitable for different environmental and urban contexts, a protocol of action to address a course of simulated and empirical testing, up to the implementation of the system and the involvement of various technical and managerial *stakeholders*.

Crossing the two variables of time and systemacity results in a vast horizon of possible cases, essential to recognize the importance, value and significance of the design operation one wishes to accomplish.

Throughout its history, design has been able to express, through various guidelines and professional fields (think of the manuals in the field of graphics, the *guidelines* of a metaprojectural nature), several distinct topographies in the vastness of the data plans from these axes. However, now we must expand the framework and produce new tropes and topoi, hybridizing and overlapping, venturing into uncharted areas, defining systems and constellations invisible to our eyes, perhaps too used to thinking about given configurations.

Such a situation is comparable to the one in which lies, since the twentieth century, ethnography, expressed with great care by ethnographist James Clifford in his fascinating essay *The Predicament of Culture* (1993):

> Ultimately my topic is a pervasive condition of off-centredness in a world of distinct meaning systems, a state of being in culture while looking at culture, a form of personal and collective self-fashioning. This predicament, not limited to scholars, writers, artists, or intellectuals, responds to the twentieth century's unprecedented overlay of traditions [...] I argue that identity, considered ethnographically, must always be mixed, relational, and inventive.

If only 'local performances on the trace of reconstructed pasts' are possible, as Clifford argues, then let us be guided by the ethnographic definition of designer from the words of a master in creating inventive grammars, as was Gianni Rodari:

> Creativity is synonymous with divergent thought that is, thinking that is capable of continually breaking the schemes of experience. A mind that is always at work is creative, a mind that always asks questions, discovers problems where others find satisfactory answers. It is a mind that prefers fluid situations where others only sense danger, a mind that is capable of making autonomous and independent judgements, that rejects everything that is codified, reshapes objects and concepts without letting itself be hindered by conformist attitudes (Rodari 1973).

4.2.2 AdvanceDesign Gradients: Consider the Ambitions of the Design

Thinking with an ethnographic spirit on the levels (we could define them perceptually like gradients) of *advance design* means understanding the importance, the complexity, and the cultural and social spectrum that you want to put into action: it would be easy to speak of 'complexity', but perhaps, without excessive abstractions, it would be fairer and less hypocritical to speak of 'ambitions' of the design. Wherever the word ambition assumes, without fear and awe, a positive and constructive sense, not directly related to self-congratulatory logic already criticized in the opening, but looking for other ways, as described by the thoughts and in the words of French philosopher Jean-Luc Nancy:

> It might be then, that the current situation of social being has to be understood in some other way than by starting from the schema of an immense, spectacular self-consumption, a schema where the truth of community is dissolved and engulfed. It might be that the phenomenon of the generalized spectacle, along with what we call the 'tele-global dimension', which accompanies it and is consubstantial with it, would reveal something else altogether [...]. Being gives itself as singular plural and, in this way, organizes itself as its own stage. We present the 'I' to ourselves, to one another, just as 'I', each time, present the 'we' to us, to one another. In this sense, there is no society without spectacle; or more precisely, there is no society without the spectacle of society (Nancy 1996).

By accepting such a scenic metaphor ("the great theatre of the world", as Descartes called it), time is perhaps, between the two symbolic variables mentioned in the previous paragraph, the one that today enables a more original and productive reflection, undervalued compared to the specific topic, already burst and also overused in the last decade. Our time is changing as the perception of space is widened and expanded to the entire planet, to the whole scene of consumption and entertainment:

> In order to prolong itself, it is said, consumer society must destroy durable things. There is no longer a slow disappearance of objects, but a 'violent loss': acquisitive enjoyment and shopping are the basis for the work of destruction of what was bought (Bodei 2009).

In fact, in fields such as art or architecture, time duration seems to be a fundamental requirement of the declared and obvious success of the creativity of its author, in design it cannot be interpreted as a guarantee or precondition to becoming a *milestone*. On the one hand, objects that were designed over a century ago are being produced (think of the Cassina collection 'The Masters', as an embezzlement towards the original clients) and, on the other hand, are the objects included in the history of design, even if they lasted only for the duration of a layout and never became available beyond that period (consider, for example, the installations/prototypes of the famous exhibition *Italy: The New Domestic Landscape* at the MoMA in New York).

Therefore, it is interesting to look not only what the market decrees and the act of consumption, since it shows inevitable variables of causation and subjectivity (until when will the pieces of the collection 'The Masters' be produced? Are the conditions imagined by the authors being respected? Without the subsequent good fortune of Italian design around the world, would the objects of *Italy: The New Domestic Landscape* have become media icons? Are they cause or effect?). But we are interested in assessing the ambition (one might say methodology) of the design, if the designer, upstream, has been able to include the temporal dimension in a conscious and programmatic form in his actions. Some examples: What is the value and duration of a product *concept* on a semantic and cultural level? Some American *dream cars* from the fifties are still famous and renowned. Meanwhile, a car design time went from an average of 7 years, until the early eighties, to the recent three, when *concept cars* are no longer a mere stylistic or formal exercise, but a planned, numbered and progressive strategy in order to develop new models respecting the tight timelines dictated by the market today. In the automotive industry, *advance design* has now become the formula by which it produces and manages innovation, and the prediction of time and compliance with development schedules form a fundamental variable, if not the generative matrix itself.

How long can a manual of *guidelines* of a product last? Information regarding the formal and geometric qualities of a product and its related families, its finishes, colours, and graphics are obviously related to the logics of consumption and market: cars are not fashion accessories, consumer electronics are not food packaging and furnishing is not publishing. However, the intelligence in designing a manual lies in its ability to anticipate trends, and also in knowing how to manage and monitor changes in the market and in taste, given also by actions and feedbacks of potential competitors.

There remains a warning of those thoughts, as designers as well as consumers, the words of the one who was able to best describe the condition of liquidity that characterizes our times, and analyzes the anxieties dictated by the perverse phenomenology embodied by our conception of time:

> A consumer's life, a life of consumption, is not limited to purchasing and possessing something. It is not even limited to the fact that we dispose of what we bought two days ago, and which until yesterday we proudly displayed. What distinguishes it more than anything else, if anything, is the fact of being in constant motion. For a society that sees in *customer satisfaction* the underlying motivation and goal to seek, the very idea of a satisfied customer is neither a motivation, nor a purpose: it is, if anything, the most terrible of threats (Bauman 2007).

4.2.3 AdvanceDesign Approaches: Define the Experimental Nature of the Design

In addition to the criteria of temporal and spatial extent, in the specific arguments that we want to deal with through the design, we cannot forget the motives, themes and forms with which one commences a process of *advance design*. In particular, the approach to the procedures we define as *advance design* can be quite different by nature and mode, implying structured and diverse levels of uniqueness and experimentation of the processes and methods (not results, here not assessable). We are fully aware of that:

> All human processing are combinatorial. That is to say, simply, that they are artefacts composed of the selection and combination of existing elements. We have seen how many problems arise from the question of the divine creation or astrophysics 'creation out of nothing' (ex nihilo). This option is not open to human beings. However, the combinations can be new, unprecedented (Steiner 2003).

We can therefore apply to *advance design* at least three large filters generated by the type of approach that we want to connote and characterize the process of the design, for simplicity we will take the examples directly from within the Politecnico di Milano.

4.2.3.1 Thematic Approach: Pioneering the Subject Addressed

Advance design can be produced and justified by focusing on and identifying specific themes, choosing to face new and unexplored issues of large-scale or high specificity, but often lacking in written information and case studies.

Today, we would not be speaking of service design, *business design, interaction design*, if there had not been an upstream action, more or less explicit or recognized, aimed at introducing an original interpretative filter to the action of the design.

An exaggerated example of the value of such an approach can be identified in the field of education: think of the numerous courses of POLI.design that have examined in depth minority and specialist themes such as Pizza Experience or Wedding Design to the limit of paradox and intellectual provocation, and perhaps of its credibility.

In this regard, I can cite again my experience, since 2006, with the Norwegian company Q-Free regarding the theme of *urban tax* and *mobility charge*, who was dealing with the rise of the socially complex issue of traffic decongestion from the centres of large metropolitan areas, facing a new framework of major political and social significance, in addition to technological *transfert* from the infrastructural sector related to highways to the one dealing with urban centres.

4.2.3.2 Instrumental Approach: Process Innovation

Advance design also deals with well-known issues and consolidated topics through resources, procedures, tools, and highly innovative and deliberately unconventional practices. Interdisciplinarity and *teamwork* are often the foundations: the processes of *cross fertilization* and *transfert* of specific and original approaches and techniques from other disciplinary fields.

In this regard, I can cite the direct experience of the practices I undertook, since 1994, during the courses at the Politecnico di Milano, alongside Giacomo Mojoli. In the Product Concept labs, we have often applied multisensory charts, typically used in food tasting, as an instrument for reading industrial objects and products, rather than using brand values.

Along with the consulting and design company Total Tool, we have instead developed, on a professional level, for Confartigianato (General Confederation of Italian Crafts), a new relational tool, called the 'Octagon of identity values of the New Craft', able to help artisans relate, in new ways, with the emerging needs of the market and with the creative flows represented by the world of design and *design-driven* companies.

4.2.3.3 Management Approach: Role Transformation

Advance design can also be seen as a self-generating process of autonomous entrepreneurship which makes, organizes and produces in an innovative way, yet directly and explicitly; it becomes an embodied form of experimentation and advanced research. The proponent often changes its institutional role and opens new ground for dialogue in a concrete and operational form.

A significant example would be the operation of Campus Point, a business incubator of the Politecnico di Milano at Lecco, where a temporary architectural structure that was set up during construction of the new campus, and became a place of research and experimentation between several local companies.

Recently, along with the law firm NCTM, I had the opportunity to jointly present a new spa and wellness concept, called Urban Spa, based on the redevelopment of former industrial sites and the strong integration of commercial activities (*wellness,* catering, sales, and services) within a single architectural volume. We defined this approach as *venture design*, since a group of legal and financial *advisors* work alongside a creative component of architects and designers upstream, not downstream, in order not only to combine business and creativity, but also to look for partners and clients who share, and therefore could support, the business potential that is later jointly defined and optimized by the corporate structure that is created *ad hoc*.

4.2.4 AdvanceDesign Authorship

One must also carefully evaluate and reflect on the nature and identity profile of the individual who promotes an action of *advance design,* since it significantly changes the meaning and expectations of value in relation to any possible recipient/purchaser.

One cannot think that the issuer of the action of *advance design* is neutral with respect to his actions and towards the modes and contents, given the strong communicative component and expressive value implied in every process of this nature.

We can distinguish certain individuals, different in nature and role. The motivation that pushes these individuals (who can obviously also move jointly with each other) and the degrees of freedom derived from it, significantly modifies the action of *advance design.*

- *Design students and young designers*. Students or young designers starting out, have a freedom to operate, or maybe even ingenuity, that facilitates the 'intelligent transgression' and rupture of given patterns, especially if and when it is inscribed in a context of given rules and parameters, whether it is a contest or a prize. However, the value of *advance design* can be attributed, without entering into further details, to the prize or contest itself, to the whole set of proposals submitted and their total combined value, panoramic and representative of the emerging thoughts of a group of young people with respect to a given topic.
- *Professional firms* If we think of the actions carried out by professional individuals and professional firms, it becomes clear that the driving factor motivating the action in *advance design* can be identified in the concepts of a formal nature rather than stylistic, or in the desire to present targeted capabilities with respect to a specific market. *Consider, for example, the concepts regularly formulated in the automotive* industry by firms such as Giugiaro or Pininfarina, instead of the *interaction design* by the American Ideo. We can, however, cite

once again the example of *Italy: The New Domestic Landscape,* if we want to travel back to 1972.

- *Schools of design and universities.* It is natural to think that design schools, public and private, and universities carry out actions of *advance design,* commissioned by particular clients and financed by public funds of various kinds or self-promoted. Education and training can be seen as significant parts within the process of *advance design,* and labs and research units can deal with projects of *advance design:* as in previous cases, teachers and advisors are the ones who speak out and comment, often for consulting and research purposes that do not necessarily have to materialize into highly figurative and iconographic scripts.
- *Magazines and websites.* In recent years, the media, primarily magazines and means of information, but also numerous websites and portals (and perhaps today we might add trade fairs), appearing more and more interested in assuming an alleged visionary function and active posture, have changed their role, by increasing their will to determine trends, phenomenon, transformations in society and consumption. Clearly, the purpose of any action of *advance design* is purely informative and communicative, although often produced through direct participation of designers and architects.
- *Corporate design centres and R&D.* Research and development centres of large companies have recently changed their strategy, starting to inform and communicate at the earliest opportunity was usually a topic of internal diffusion only, not to mention trade secrets, alleged or otherwise. If the automotive world were to break a record with such a strategy, we cannot but mention the one recently introduced by Stefano Marzano with his arrival at Philips Design, beginning with the well-known Vision of the Future.

We can, therefore, conclude without dwelling on what would be a very interesting survey if completed, or rather, a true history of design parallel to the one that is best-known, that in *advance design* authorial diversity is what structures and expands its definition and phenomenology. Certainly we have not listed all the possible individuals, we have not traced the possible interactions, but it is clear how the outlook is structured, rich, complex, and constantly evolving and becoming.

4.2.5 AdvanceDesign Principles

Great potential of political and qualitative nature is embedded in *advance design,* linked to the desire to develop certain principles of the design discipline. Innovation is not easy, indeed it is unnatural by definition: in innovating the professionalism of innovators will inevitably inscribe itself in the same theorem. It is perhaps more about motivation, principles, moral and ethical issues than market factors or professional qualifications. Let us try to enucleate them, as if they were the first points of a hypothetical manifest to be written only through proactive and concrete steps of *advance design.*

4.2.5.1 Epigenetic of Design

Design is an ongoing, open process, which is inevitably transformed in its becoming. Its shapes and metabolisms must be designed, breaths and rhythms must be heard, including cycles and intervals, anticipate the stages of transformation, and sustain the stages of potential development.

Traditionally, one always spoke of industrial design, but perhaps we should move away from the artificial and self-reported flavour that the word implies, and instead, assume an agricultural attitude, where the seasons and climates, the unexpected as well as the hearing, tenacity and intuition are the winning factors within an organic and pan vision of the processes, both micro and macro.

Advance design takes the epigenetic dimension of the design, the time value of adaptability of the transformation of the world and its constant turmoil, local and global, tangible and intangible, cultural and economic (Ceppi 2010, 2011).

4.2.5.2 Cross Fertilization

A design is hybridization of knowledge, crossroads of expertise, a dialogue between diverse and specific cultures that positively pollinate and contaminate each other. A designer must speak different 'languages', communicate and interface knowledge often incommunicable: research, production, marketing, communication, distribution, etc.

Only cooperation between different worlds, dialogue and interpolation, a desire to compare codes and knowledge can produce an in-depth and detailed exploration of the new, starting indeed from the diversity and differences, and taking each other into account.

Advance design starts from the interference of processes and knowledge, from the cognitive game of exchange and interaction, through an active and conscious orchestration, although not without intellectual and methodological provocation and potential discontinuities.

4.2.5.3 Venture Approach

Design is often the anticipation of the market through the understanding of the models of social and real economy, of the logic of production and supply chains, and careful analysis of the value chain and its articulation with respect to different *stakeholders.*

This does not mean challenging a possible briefing in a radical manner, nor giving up the idea of the company as a mainstay client, let alone think about being 'self-entrepreneurs' or 'doing business', substituting the lack or longing with ideal and perfect clients.

Assuming that all these reactions are justified (there would be numerous citable examples), we propose a new axiom: that along with the 'commodity-form' one might imagine a new 'enterprise-form', putting also directly into play the proposing parties.

Advance design proposes, anticipates, if necessary embodies, not only as a product and system solution, but also as complimentary entrepreneurial models, economically sustainable systems that know how to diversify or integrate with respect to existing business practices, rather than propose new *tout-court.*

4.2.5.4 Virginity

In recent decades, design has branched and structured its phenomenology into nominal forms of specialized design (*transportation design, interaction design, strategic design,* etc.), and by defining new disciplinary fields and approaches on both a professional and educational level. The process is clearly still growing, and design is taking possession of lands and is putting his own endings to mark the numerous territories, whether it be investigative methodology rather than commodity sectors, transient or permanent.

Advance design continues somehow this disciplinary structuring phase of design in various fields of knowledge and making, but in some way also seeks interstitial spaces, assembles fragments, reverses routes, proposes challenges, driven more by the spirit of exploration or from the study of the authentic, than from the spirit of conquest. *Advance design* should be more supported by processes than by classificatory and nominal anxiety: with the risk of renaming the 'known' with a new suffix, an authentic and virgin 'not-known' is preferable, maybe even vertigo.

4.2.5.5 Participative Storytelling

In recent years, especially from experiences in Northern Europe and in Anglo-Saxon culture, there has been much talk about 'participative design' and about creative and decision-making processes that involve the end users within the design and the definition of the requirements of the product/service. In addition, social concepts and practices such as distributed economics or intermediation, are changing not only the design of products, but also the entire production chain, by connecting production and distribution in a new and direct way, with an approach that starts precisely from below.

Personally, I do not believe that design can be a strongly democratic activity with an exclusively social impact, because there is always the mark of a decision maker/coordinator, of a process handler, that I think is wrong to neglect or even want to cancel: I believe that participation serves as an instrument of dialogue and sharing, of analysis and wider understanding of phenomena, and that, in such sense, it is an instrument and a state of absolute relevance in the practise of *advance design*. However, there remains a middle and not an end, whose value is played more on a narrative level (*storytelling*) than in concrete solutions, which then need a final decision maker, an actor who has the task of giving a shape, a smell, a voice, elements outlined in grammars set on things, perhaps even a bit of courage and poetry.

4.2.5.6 Sustainable Sensoriality

During preparation of the conference 'Slow+Design'[1] we used for the first time, alongside Mojoli, the term 'sustainable sensoriality', wanting to celebrate and show the importance of combining in a happy and harmonious way the dimension of sustainability and the practices of eco-design with the need for products that can be enjoyed sensorially, the result of careful design evaluation from an aesthetic level and from the final sensory experience (see Footnote 1). The fact of combining the principles already expressed in the famous motto of Slow Food in relation to the 'Good, fair, clean' and translate them into the design and definition of industrial goods and services, seems as an important value for any future activity of *advance design*. Criteria such as traceability and territoriality, understanding chain values and relations, the ethics of individuals and relationships underlying manual production or industrial processes, the evaluation of the sensory and experiential quality of a product, are now considered basic requirements of any design.

To conclude this open and evolving list, we can only quote again the words of Steiner (2003):

> It is scientific discovery and technological invention which will, more and more, marshal our sense of social history and of the idiom appropriate to that history. It is already possible to find such elegance, such aesthetic adventures in architecture, in industrial design. These are the synapses between the arts, in some kind of traditional sense, the algebra of the engineer and the virtuosity of the craftsman (Cellini would have been delighted with a Ferrari). In this symbiosis, the partitions between what is created and what is invented have lost definition. Having listened to Duchamp, Brancusi embodies in his sculptures the dancing curves of a propeller. One senses that, in the arts, this will be the next chapter.

We hope that our next chapter of life is made of synapses and symbiosis, of elegance and aesthetic adventure, listening and creating, and putting the past to sit in the future, just like Cellini in a Ferraricar.

[1] The conference was sponsored and organized by the Politecnico di Milano, European Institute of Design, Domus Academy and the University of Gastronomic Sciences in Pollenzo, in 2006, with the support of the Province of Milan. About the concept of 'sustainable sensoriality', please refer to the concerning documents (www.dis.polimi.it/manzini-papers/slow+design_background.pdf).

References

Baudrillard, J.: Il sogno della merce. Lupetti, Milano (1987)
Bauman, Z.: Homo consumens: lo sciame inquieto dei consumatori e la miseria degli esclusi. Erickson, Gardolo (2007)
Bauman, Z.: The Art of Life. Polity, Cambridge (2008)
Bodei, R.: La vita delle cose. Laterza, Roma-Bari (2009)
Ceppi, G.: Epigenesi del design. Aracne edizioni, Roma (2010)
Ceppi, G. Design Storytelling, Fusto Lupetti Editore, Milano (2011)
Clifford, J.: The Predicament of Culture. Twentieth-Century Ethnography, Literature, Art. University of California Press, Berkeley (1988)
Nancy, J.L.: Être singulier pluriel. Galilé, Paris (1996)
Rodari, G.: Grammatica della fantasia: introduzione all'arte di inventare storie. Einaudi, Torino (1973)
Sennett, R.: The Craftsman. Allen Lane, London (2008)
Steiner, G.: Grammar of Creation. Faber & faber, London (2001)

Chapter 5
The Role of Humanistic Disciplines in a Pedagogy of *AdvanceDesign*

Antonella Penati

> *During revolutions, scientists see new, different things, even when they look with traditional instruments in the same directions theyd already looked in.*
>
> Thomas Samuel Kuhn

5.1 Design in the Contemporary Scenario

Third wave, megatrend, era of all things discontinuous, era of the post-market, neo-capitalism, post-capitalism, post-industrial, super-industrial, neo-artisan, neo-fordism, post-fordism, economy of access, economy of creativity, economy of learning. These are just some of the terms taken from the vocabulary of sociologists and economists, with whom an attempt has been made at giving a face to change, enhancing, often in a simplistic manner, its radical nature.

In managerial disciplines, use is made of the term turbulence or even hyper-turbulence, to describe the capacity, speed and multiplicity of the points that trigger the dynamogenous processes that would characterize our time.

Product prices change, consumer attitudes change as do the nature, times, methods and places of production and distribution, but most importantly, change is seen in the fundamental data concerning technology, the political–social environment and the cultural systems of reference. At the same time, there are changes in values, lifestyles, behaviours and consumer scenarios. The reference context and choice of social and economic players are no longer described as stable and certain *data*. The places of incubation of innovation are widespread and the single enterprise, with its R&D centres, is no longer the sole point of reference. In the same way, the dynamics produced by technological innovation take on complex inter-sectorial trajectories which are not always predictable and spread rapidly, also

A. Penati (✉)
Dipartimento di Design, Politecnico di Milano, Via Durando 38A, 20158 Milano, Italy
e-mail: antonella.penati@polimi.it

M. Celi (ed.), *Advanced Design Cultures*, DOI 10.1007/978-3-319-08602-6_5

becoming quickly obsolescent. Products no longer respond to simple functional needs but break through the sphere of emotions, aesthetic and sensorial sensitivities and of values in terms of the culture of individuals.

The ability to adapt to change, contextualized in terms of continuous learning seems, in this context, to be a prerogative required of single workers, including those with routine jobs, who must however demonstrate logical research skills (Sasso 1999; Nonaka and Takeuchi 1997) which implicate the refinement of new cognitive instruments.

This happens even more so in the professional and business worlds, worlds which are called upon also to express capacities of vision, capacities to interpret current trends, not only to describe possible scenarios but also to project them towards the definition of socially useful and desirable futures.

Pre-vision, conceptualized as a special way of thinking, organizing and acting—a structured field of action in which purposes, resources and restrictions are visible (Ceriani 1996)—allows the ordering and organization of information, knowledge and ideas and their finalization in the transformation of existing situations into desired situations (Simon 1969) characteristic of design, in so far as it is a cognitive and operative instrument of operation and alteration of the context, which has been characterized for some time by adjectives such as 'strategic design' or 'advance design', highlighting the capacity for anticipation together with the tendency towards systemic action.

We know that in every design-related action there are closely connected activities of configuration—i.e. giving elements a conceptual and physical placement such as to give them a structure that is functional to the expected result—and the activity of prefiguring, i.e. the capacity to anticipate the possible expectations in the design action.

Both these polarities of the design sphere are subject to considerable stress when configuration—an activity which can often be traced back to creating an order, giving sense to and recombining given elements to create innovation—is compared with an explosion of complexities that touches products at structural, formal, performance, technological and productive levels, making resources and restrictions that have always formed the tracks capable of orienting design actions uncertain. Furthermore, the dynamics of continuous change of contextual data pose problematic questions relating to design, and to innovation, which are transformed from actions aimed at pursuing sufficiently defined aims established in time into subordinate actions at progressively growing levels of uncertainty.

It not only happens in all decision-making activities, but also in design, a factual dimension and a cognitive dimension are indissolubly united. This second part takes on an extremely important role in the face of the complexity of the technological, social, cultural and political sphere and in the new awareness of the entwinements and deep contaminations between these different systems.

5.2 *AdvanceDesign* and Radical Innovation: Problems of Conceptualisation

In the essays in the previous text, *advance design* is not subjected so much to the presentation of proposals of definition—which often disguise much more complex situations and reduce them considerably at terminological level—as indications relating to contexts, modes and the nature of its practices.

The capacity to reformulate design issues rather than responding to a brief consideration; the passage from design as an act of individual thought to participated design and which is submerged in the social context to interpret it; design as an instrument to produce new value and strengthen the identity of the players more than a means to reach mere commercial purposes are just some of the introductory elements—summarized briefly here—that are mentioned in Giulio Ceppi's essay as the foundation for the practice of *advance design* in which the timescale of the design process and the capacity to assess the different levels of systemic implication what accompany every single design intervention play a central role even though carried out at product or single component level.

Of the prerogatives attributed to *advance design* and that distinguish it from the design practices aimed at more detailed operations, in this chapter, there are two particularly important issues that I would like to look at before proposing some problems relating to a pedagogy of *advance design*.

The design intervention actions that place in the advancement on the timescale and attention to repercussions, effects and epiphenomena that the single design act generates on a broad scale (social, cultural, economy, technological, environmental, etc.), bring to mind to terms which, in the literature of the theories and culture of innovation are recurrent and possess a strong critical element: I'm referring particularly to the character of *radical innovation* which is attributed to certain design intervention actions and to the *systemic dimension* that these can assume.

In literature, the term 'radical innovative' appears at a time when there is a need to refer to new forms of production which seem to proceed 'in leaps and bounds' rather than following linear or incremental development.

While the issue has found its own place in informative essays, particularly linked to an effective descriptive impact that this term is able to produce, it should also be pointed out that it has often been accompanied by considerable problems particularly in theoretical texts that pay closer attention to the dynamics that differentiate it from other forms of innovation.

We can also say that, while a rather consistent amount of thought on incremental innovation has been developed, within the scope of different studies, contributing to form the methods that give life to innovation, we have few references to forms of radical innovation.

On the contrary, despite the diversity of the disciplinary spheres that have looked at innovation, it seems that the dominant feature is the fact that, in order to be consolidated, something new has to present the need to build different kinds of links with things that already exist.

So, in the field of technological innovation, the reference to '*path dependent*' methods has been developed to explain how the path travelled forms an element capable of influencing the subsequent passages of an innovative process (David 1985). Remaining within this area of study, and particularly within evolutionary theories which refer to the dynamics that dominate the biological world, it is presumed that the production of innovation often originates from *bricolage* operations that lead to the production of something new by reusing, transferring and missing existing elements (Ceruti 1995). Schumpeter began his *Business cycles* by sustaining that the majority of innovations are the product of new combinations of old knowledge (Schumpeter 1939).

In business theories, we try to explain the introduction of innovation as something that penetrates the business context according to logics that are closely related to what the business already possesses. On this matter, Edith Penrose has tried to explain the processes of business growth, via diversification, starting from the theory that the resources that make up the nucleus of skills of a company substantially determine the range of business action, or its ability to move away, through innovation, from what it knows and how to do (Penrose 1959). In this case too, it is assumed that the new combination of excess resources is the spring that triggers innovation.

In an essay which aims to analyse the dynamics of the introduction of the major technologies in companies that move within the spheres of traditional technologies, Petroni (2000) has developed an interesting reflection on the basis of which the process of intromission of new into old takes place according to a 'cone-shaped' model. These businesses are capable of absorbing something new only within a process of specialization of the old. The more specialization advances, through processes of furthering of technical-scientific knowledge already possessed, the more the range of diversification of knowledge expands. In other words, innovation does not come from outside but is generated through the creations of complexes and analysis of internal lines that expand into contiguous segments of knowledge. This model is very similar to those developed in the world of research, which envisage two ways of attacking opposing but complementary problems: the generalist and the specialist model. During the research phases, both models converge in the production of something new but the capacity to generalize is always linked to forms of specialization, or to a remarkable capacity to manage the elements to be recombined. In this case too, 'the new' is generated largely due to 'the old'.

Within the theory of systems, we know how important the mechanisms of morphostasis and morphogenesis have become, offsetting each other to allow the new to avoid replacing its predecessors but to be absorbed into the system. In this case too, the processes of morphostasis perfectly represent the way in which the new comes to terms with what is already there (Emery 1969). In evolutionary business theories this mechanism has been clearly analyzed by Nelson and Winter, who have attributed a role of fundamental importance to organizational routines as forms of memory of organization that would allow concentration on the new when the old got sedimented, penetrating into the action of the organization, through crystallization into fixed procedures (Nelson and Winter 1982).

It seems obvious that, while we have a rather important series of theoretic references which can help us to understand the process of the new when this takes on incremental forms, we have inadequate references to interpret the methods of production and the forms of the radically new.
This plan lacks a reflection that supplied models of reference for design, such as activities to solve problems and for innovation.

Here we also refer to one of the critical elements that recur in the definition of the genesis of processes of radical innovation, which are central, in my opinion, to the purposes of what follows: the character of 'breakage' produced by an innovative action is often referred to the particular skills of intuition and inventive genius of a subject or of a single business when, vice versa, a closer analysis shows how especially *performances* that take place during the diffusion of innovation tangibly define the characteristics of the impact more than those that take place during the moment of concept.

Paying attention to the dynamics of *post innovation performance* (Georghiou et al. 1986) places us in front of certain important theoretic nodes:

(1) attention shifts from the search for the actual moment when the innovative shift takes place to concentrate on the process-based dimension that accompanies the emerging and diffusion of innovation with some of its characteristics of rupture and radicalness;
(2) interest in the search for characteristics that make the creative skills of the single inventor exceptional is mitigated, paying more attention to the characteristics of the inventive, innovative and diffusive context in their technical, social and economic tangibility and the cultural climate.

In the first context, those studies that analyzed the so-called 'technological corridors' in which the initial invention is subjected—almost in a sort of 'via crucis'—(Maldonado 1987) to productive, technical and functional checks which then take a product or a device onto the market in a substantially more mature form than that originally generated by the minds of the inventors of the research centres, were exemplary. But numerous incremental passages along the path of maturity of the product are also generated by social restrictions and not only by those linked to use. Just think of the importance attributed to the forms of innovation linked to the contributions of the end user's experience, the latter being a player capable of producing consistent innovations. The consumer electronics sector, as opposed to the worlds of extreme experimentation such as medicine, sports, etc., are, in this case, references of outstanding importance. Lastly, the market, being the place where economic convenience and social instances (tastes, needs, expectations, etc.) are manifested, forms another filter capable of amplifying, mitigating and modifying the capacity and the nature or an innovation.

In other words, however radical the results of an innovative action might seem, we find ourselves facing a complex process and not a detailed event capable of consistently—and radically—modifying what already exists. Whether this refers to a product, a process, an organization system or a customary or social practice.

On a second level, despite continuing to receive considerable interest, the studies originating mainly from the world of cognitive psychology that has very helpfully probed the processes of individual creativity, encoding the various passages of exploration and reconstruction of a problem in view of innovative solutions (subjects on which we will concentrate further ahead), the theoretic attention has now shifted to collective forms of creativity, including once again, particularly thanks to the contribution of social psychology and studies coming from the sociological world, processes of individual innovation, within the broadest sphere of groups, communities and society as a whole.

In terms of design practices, when we refer to the term *advance design* and its prerogative and tendency to create those forms of cognitive, function, formal caesura, and also of sense, with respect to what already exists—which we have just attributed to the processes of radical innovation— we find ourselves faced with these same problems.

Because the majority of studies of the creativity of the designer have clearly highlighted that it consists of a sort of multiphase structure which, far from being an act of instant intuition, purely metaphoric, bases its strength in a substrate of complex materials (data collected, technical and economic restrictions, aesthetic sensitivities, cultural variables, verbal or visual memories, sensorial data, etc.) and of micro choices in succession that close certain alternatives and open up others (De Fusco 2008).

Because many of the innovations introduced, including those of extensive capacity, seem to make reference to forms of transferral of scale, context and sector, to forms of reinvention (Rogers 1983); to forms of reuse and combinations of existing elements (Ceruti 1995); to forms of decomposition and structural recomposition of technical, material or cognitive elements; to forms of logic and analogic relationship with what already exists rather than to authentic ruptures lacking lines of ascendance.

The innovations of the strongest caesura with what already exists come from acts of logic and systematic intelligence that are closer to a process of cognitive reconfiguration than to a process of cognitive rupture or an absence of logic or rules.

And lastly, because, while it's true that there is a strong link between creativity and innovation, we must not forget, as highlighted by Legrenzi, that in the first case, we have to face up to the skills of single individuals, while when we talk about innovation, we are faced with a social phenomenon that has to be investigated as such (Legrenzi 2005).

5.3 *AdvanceDesign* and Systemic Innovation

The radical nature of the innovative processes, despite being mitigated by the many critical objections that this term has received and despite its inadequacy in grasping certain important elements of the production dynamics of the new, has the advantage of having opened a considerable amount of interest in various spheres of study of the cognitive behaviours of individuals and the various forms of creative thought on which we will focus farther ahead.

The use of the term 'radical' and 'discontinuous' has, however, produced in the theories on innovation a sort of disinterest in looking at the innovative phenomenon in its process and a closure of the investigation of the technical, scientific and social context in which this process takes place.

Looking at innovations in their process-related dimension means shedding light on many factors that intervene along the development path to mould their characteristics and define the directions, sometimes causing inventions to run aground or change, even substantially, the initial characteristics and functions, the performances and the applicative possibilities.

Focusing attention on the process-related dimension of change means following the metamorphosis of a technology or a product during its lifecycle. The slow transformation that leads an artefact to pass from the invention phase to the diffusion phase cannot be analyzed making exclusive reference to the dynamics within a narrow technological context. The important complementary relationships with products, processes or, more generally, with different technological contexts and at different stages of maturity, show us that inventions very rarely operate in isolated conditions. It's quite common for the success of an innovation to depend on the availability of goods, instruments and complementary products for which it is sometimes necessary for additional innovation processes to be activated in other sectors.

Another important passage in the arrangement of the processes of innovation induced by design activities consists in opening the investigation of the reading of the complex context made up of resources, which are also restrictions and opportunities for innovation, such as those of an economic-productive nature, such as scientific knowledge, the cultural climate, finances available, institutional and political interests, legislative restrictions, etc., but also the interests of individual subjects, groups, bodies and institutions that select and sustain certain technical options and eliminate others. The context, therefore, is also—and above all—a social context that feeds needs and expectations, creates technological problems but also the possibility and methodology with which to solve them (Penati 1999).

The entrance of society as a combination of stakeholders is possibly the least innocuous entrance in terms of the studies of innovation. It makes it possible to focus less attention on the mythological character with which certain innovations have been analyzed and bypass the biographic reading of single inventors blessed with talent and personal ability to think of innovation as a socially built process.

And just like stories that reveal the human condition in the singularity of an individual (Morin 1999), in the most recent stories of innovation, single inventive acts, rather than being read as the exceptionality of a genial mind, are charged with the knowledge and expectations of an entire moment in history.

So we can say, with Legrenzi, that 'innovating equals recognizing in an event something that no one had seen before but which many would have liked to have seen' (Legrenzi 2005).

The appearance of the same inventions at the same time but in different geographical places also happens—just think of the sewing machine, the light bulb and all those inventions that have aroused so much interest in the histories of technique, in which we sense the anxiety to trace the origins, dating and explaining, attributing paternity of the invention to one inventor or another—but finds an obvious explanation in the presupposition that 'ideas are in the air' because they are produced by cultural curiosity, by the scientific climate with technical explorations and by 'technical ideals' (Feibleman 1961), by political desires, by aesthetic, expressive and linguistic sensitivities and needs of the different social systems.

And once again the complexity of the social sphere dominates the scene in explaining the nature, direction and radicalism of a change and the inventor is, at best, the person who interprets the instances of a context to the extent in which he is fully involved, absorbing and interpreting 'the climate and cognitive style' in use, the behaviours and the consolidated habits that form a paradigm of reference—to use the term with which Kuhn defined periods of 'normal science' and periods of 'extraordinary science' in which a leap in paradigm is produced—(Kuhn 1962), but at the same time it possesses antennae capable of sensing the warnings of change and the signals of a reorientation of the shared feeling that it is able to guide through its design-related action.

Seeing before others what others cannot see but expect to see is one of the prerogatives that, at the beginning of this text, we attributed to the practices of *advance design.*

Capacity for anticipation, which produces real innovation only insofar as it is able to grasp and interpret the instances of the numerous subjective and collective, private and public players that it is able to involve—ideally or politically—in the process of elaboration of the new, mediating the different instances.

But also the capacity to grasp the complex combination of actions, repercussions, effects and processes which, starting from the innovation proposed, involve the material, social, economic and cultural spheres, etc.

The capacity to grasp the dominant features of a cultural, stylistic, technological, ethical paradigm—just think of the environmental question, for example, as a strong device for focusing interest, practices, design logics and the innovation of contemporaneity—interpreting them.

Lastly, the capacity to build instruments, places, methods of sharing, orientation, conversion of the systems of value, of lifestyles, of consolidated behaviours when the impact of design-related proposals is such as to redirect the shared sense and has to be accepted within new social imageries.

Considering *advance design* as a practice of design which enhances the dimension of systemic intervention on the interconnected spheres of technique, society and culture have important effects on the forms and contents of a pedagogy of design. I'll take a closer look at this specific point later.

5.4 The Contribution of Humanistic Disciplines to a Pedagogy of Design

There is plenty of theoretic evidence that allows us to place the practice of design in a strategic point of the coevolution processes that link social and cultural systems with the system of what is artificial and its economies.

This is one of the main reasons for legitimizing the construction of didactic paths which, at international level, since they first appeared, haven't simply limited the list of disciplines in the teachings of design culture and in those that fall within the sphere of technical intervention, but also draw on different kinds of theoretic-critical knowledge which starts from man as a cultural subject.

This is why, in the composition of the formative path, extensive space is given to the humanistic disciplines and the understanding of the dynamics that relate to the cultural sphere and the micro-scale where the subject emerges as user and person (Pizzocaro 2009), bearer of needs linked to the body, to sensitivity, to cognitive skills and to individual culture, as well as to the scale of social macro phenomena where the process of definition of the symbolic, expressive and linguistic aspects that transform tangible artefacts into cultural artefacts takes place. This is the element of profound distance that separates design from design-related practices that take place in the engineering sphere.

In the economy of this text, I have drawn up a list of some of these important disciplines, summarizing certain elements that make an important contribution, at different levels, to the didactics of design:

- *Historical-critical disciplines* which from the History of design to the History of architecture, modern and contemporary art, graphic design and communication systems, science and techniques contribute to introducing the theoretic and operational instruments that can be traced to artistic practices, restoring the method for analytical study and the critical comprehension of the intrinsic qualities of a work or of an artefact, expressed in its tangible form with the specific technical modalities and linguistic values in relation to the use and its meaning in the particular economic, cultural, social or scientific context in which it was produced. In other words, they supply the theoretic-critical foundations needed to place in a historical-evolutionary key of the expressive and formal languages, techniques, materials and the methods of transformation historically determined, the currents, styles and fashions, practices, policies and institutions, figures of importance and schools, which make the tangible and artistic culture of a period the privileged key to reading cultural systems and their evolutionary contexts;
- *Human sciences*—particularly psychology—which introduce the foundations of the perception (visual, haptic and synesthetic), of the chromatic component, light, morphology and the processes of morphogenesis which, together with the historical and artistic disciplines and semiotic disciplines, help strengthen a theory of the fundamental form in the processes of understanding the dynamics

of configuration, compositional balance and practical and conceptual manipulation of the elements that intervene in the processes of formal synthesis, etc.;

- *Ergonomic disciplines* which help in the understanding of the behavioural, cognitive and interactive component with physical and communicative artefacts which allow a reading of the world of users from the point of view of dimensional and operational conditioning, of usability, fruition; of comprehension, which has witnessed a clear evolution—in line with the main theoretic acquisitions present in the cultural debate on design around the role of the user and the project of design—of the disciplines included in the didactic framework which from the traditional studies on ergonomics and then on cognitive ergonomics, have progressed to the current 'studies on the user' which systematically embrace all the components—economic, psychological, social, relational, cultural, physical, behavioural, etc. and allow the formulation of certain explanations related to the processes of interaction between subjects and also between subjects and objects;
- *Theories of languages and signing systems* as the philosophy of language, semiotics, the theory and sociology of communication which supply the conceptual apparatus which is vital to a coherent and systematic approach to the variety of phenomena of languages in their plurality and diversification, which help in the design of goods, services, environments, communication, etc. with particular attention to problems linked to the sense of things, to communication through objects, to the meanings of spatial articulations, proxemics, etc. to the symbolic meanings of objects, to the processes of construction of social imaginaries and shares values, etc.;
- *Social and cultural anthropology and ethnographic studies*, for contributions to the construction of the systems of meaning, sense, practices, beliefs, uses, customs, rituals and myths and of all those skills, behaviours and habits that depend closely on the groups and social and cultural contexts that they belong to, in which the needs and necessities of the consumption and use of objects and relative behaviours are also built, and in which real or symbolic meanings are attributed to the possession of objects, to their formal characteristics, etc.;
- *Economic and social disciplines* which, besides allowing the understanding of the economic mechanisms that regulate the market, introduce social and cultural factors that lead to the formation of needs and demand, as well as the study of social changes and the main trends in contemporary society in close relation to the cultural models of reference.

However, within the scope of this intervention, I wanted to focus on the area that we can call *innovation cultures,* an area which, far from forming a discipline, crosses every one of the disciplinary contexts that we have listed, taking on the assumptions originating from the world of history, philosophy, science and technology, sociological and economic cultures, etc., as its own to re-elaborate some of them and reveal new conceptualisations and problems linked to the dynamics of innovation and their contribution in the educational sphere. I'll look at some of these further ahead.

I don't wish to forget that the *pedagogy of design* has drawn attention to many new issues relating to the practice of design which takes place within scenarios characterized by continuous innovative dynamics, which have important effects on the educational process. For example, questions are posed in relation to the different mechanisms that govern the sedimentation processes of knowledge when its dominant character seems to be extreme perishability, and lots of theoretic space is offered to the enhancement of the value of the capacity to identify places of availability of and access to knowledge, detracting importance from the cognitive mechanisms that regulate the process of a form of knowledge (Amietta 2000).

These considerations have important effects on the practices that form a design culture and prove to be interesting in tackling both the issue of individual creativity and that of innovation as a social dynamic. We are going to look at these two sides of the problem in the following paragraph.

5.5 The Contribution of *Innovation Cultures* to a Pedagogy of *AdvanceDesign*

5.5.1 The Studies Around Individual Creativity

If we look at the first issue, individual creativity, the attention to forms of radical innovation, in the sense and within the limits that we have already defined, has the advantage of bringing certain theoretic problems in the pedagogy of creative action to the fore.

Within the field of *cognitive sciences*—which we are referring to, here, no longer for the extensive contribution that they make to the training of designers but with specific reference to the studies on the processes that oversee the activities of invention and discovery—the heuristic thought that takes place in the solution of problems, as practice that produces innovation, has always been theorized as thought that makes reference to what we already know. Within the scope of these studies, distinction has been made between programmed decisions—in which relatively simple psychological processes such as habit and memory are used—and non-programmed decisions—in which complex learning processes on which psychological sciences have supplied considerable progress are brought into play. One thing which is obvious, thanks to these studies, is that the solution to problems always starts from comparing the new problem with others that have already been tackled and have had a positive solution, because this way of proceeding allows savings on cognitive energy. 'The search for similarities or differences, the comparison with our past experience and with categories of problems that have already been solved, forms a fundamental part of solving new problems' (Simon 1960). The least tiring way available to us to solve a problem is that of referring to our experience and establishing whether it can be considered similar to another problem to which we already have a solution, searching in our memory for the solutions applied in previous cases.

The nodal point of the strategy based on searching for a similar problem which has already been solved lies in having a good definition of the problem and a valid management of the criteria of analogy and similarity with the same structural, formal, functional and performance-related basis (Basalla 1991)[1] indispensable to choose solutions that are effective at dealing with the problem in hand (Bara 1999). One of the most obvious advantages of experience, as a form of acquisition of knowledge, values, contextual information and specialized skill capable of supplying a reference framework for the assessment and assimilation of the new takes us back to the fact that it provides a historic view (Davenport and Prusak 1997) through which to observe and understand new situations and new events.

All these references have been used within the scope of design and education to design where, from my point of view, there is still a lack of a serious study on the forms of creativity that take place there and on their nature. Even where an attempt has been made at articulating a reasoning around the production of innovation, from a theoretic point of view, investigation of the real processes behind creativity has been neglected. In these studies, at best, a contraposition between fantasy and rationality, creativity and method as two ways, and methods classed as antithetic, has been assumed. Around this node, way back in the 1960s, Maldonado intervened to highlight the dialectic and not antithetic relationship that is set up in design processes between method and creativity (Maldonado 1974).

Similarly, there is the antithesis between accumulating knowledge through the development of mnemonic skills and eliminating the overload of information by cancelling data and information (Oliverio 2003); like the antithesis between experience and intuition, between following encoded rules and experimenting in the absence of rules, etc. are all complementary methods of the design process that belong to the same area of action, supporting each other and integrating (Morin 1999) in generating the design result. In other words, the position of those who saw creativity as an anti-method has been abandoned, in the same way that there has been a shift away from the hypothesis of imagining that innovations that subvert things that already exist derive from processes of extreme creativity, from the essence of rules, from the censure of the past, from pure intuition unhinged from experience.

I think that, with regard to methods of production of things that are new, due to the radicalness of which it is often sustained that the intervention of pure creativity, intuition and a lack of rules is dominant, it is necessary to reopen research into the ways in which creativity and method are compared. In particular, we need new theoretic explorations that enable us to arrange forms of creativity that take place

---

[1] George Basalla, *L'evoluzione della tecnologia*, Rizzoli, Milan 1991. The examples used in support of this hypothesis are quite well known. Basalla, for example, looks at certain cases like the transistor the original 'contact point' form was the product of an imitation of the thermionic valve, or like the slavish structural imitation of the first electric systems that completely followed gas distribution systems, or like the first radios, which limited their functional potential simply because they were born to replace the telegraph on which they had modelled the performance-related possibilities of the new appliances.

within super-restricted socio-technical systems (Johnson–Laird 1988). Some studies have investigated the influence that the objectives of a design action and the restrictions of context produce on creative acts (Skolimowski 1966). In other cases, the process-related character is highlighted. For example, the spheres of the philosophy of technique propose the model of *cumulative synthesis* developed by Abbot Payson Usher in his *History of mechanical invention* (Usher 1954), in which it seems clear that, in the heuristic processes that arise within strongly restricted systems—such are the processes that act on technologies and knowledge that take place within social systems that finalize the design action, creativity never coincides with an instant inventive act—as may happen in processes linked to forms of art. We are always faced, instead, with a multistage architecture, a process made up of subsequent steps, each of which re-elaborates and recombines—also via intuitive actions—data, knowledge and information sedimented during the process, compared with the objectives and needs of the context.

Again, in cognitive psychology studies, the substantial difficulty experienced by human beings in solving new problems outside the logical processes trained through experience and acquired mental schemes has been revealed.

Interesting on this matter is the didactic experience of Paulo Belo Reyes, in the recent congress on education in the field of design held in Turin in November 2011, where he submitted a brief to a class of students for the design of a type of fantastic and imaginary object, the students followed a sequence of strongly structured logical operations linked to past experiences and to real data sometimes re-elaborated in metaphoric terms taken as reference (Reyes 2011).

Other studies have highlighted how the logical sequences in design follow linear and sequential structures (Simon 1984)[2] and in other points produce logic gaps. In semiotics, these gaps, which belong neither to inductive processes nor to those of deduction, have been defined as abductive, indicating a re-elaboration of the process of knowledge capable of introducing new meanings that restructure the problematic starting area.

Lastly, in studies on design practices, the tendency not to isolate the different aspects of the design problem to then recompose them, but to proceed by tackling several aspects of the same problem at the same time, using different areas of skill in the same moment, as the prevalent way of reasoning has been highlighted. Jorge Frascara has defined this way of concomitant action as *rhizomatic* logic (Frascara 2000). A logic which we find described in another way by Schon assuming a capacity for thought not so much aimed at the optimal result and maximum efficiency—as happens in engineering design—nor to the progressive reduction of the complexity of the elements at stake, but conveying value to subjective reflection and interpretation as one of the fundamental process elements.

---

[2] The substantial restrictions that lead us to tackle issues sequentially, one at a time and one after another, have been extensively investigated by Herbert Simon, who unveiled at least three orders of restriction: limits in attention; limits linked to the capacity to tackle multiple values, limits liked to forms of rationalism limited in situations of uncertainty. On this matter, see Herbert, Simon, *La ragione nelle vicende umane*, Il Mulino, Bologna 1984.

I have given extensive space to certain studies on creative thought, both to demonstrate the flourishing of reflections around this terms which is often used superficially, but also to highlight the areas still to be probed to supply useful interpretations to the dynamics of design as a tool for innovation.

5.5.2 Studies Around Innovation as a Social Process

All the theories and reflections around creativity tend to attribute more or less exclusive attention to the single individual or, at most, to small groups—of designers, engineers, researchers, inventors—on whose genius the extent, the success and the capacity of an innovation to take hold in the social, economic and technological system depend.

Many studies, vice versa, have highlighted how the same individual creativity that is unleashed in the inventive moment is nothing other than the result of the influence of broader cultural and social forces capable of producing continuous flows of ideas, hypotheses, theories that circulate and permeate certain specific problematic contexts—the world of science, technology, design—and the relative communities of reference.

These are interpretative advancements that we find particularly in *sociological studies* and even more specifically in the *sociology of science and technique* which has brought clearly into focus the particular methods of relation and action of individuals who belong to communities that produce new knowledge.

But, as far as the places of incubation of change are concerned—when many of passable inventions that become innovations first see the light—it is less and less possible to encode them, because they are present in numerous points of the social system, particularly in the moment of dissemination and use of the invention that makes its debut and mobilizes a number of individual and collective players, each with their own specific interests.

It is from sociological, anthropological and ethnographic studies that we borrow new concepts of *socio-technical system*; of *community of practices*; of *innovation as a socially built process* or *innovation as a situated process* depending on the context, etc. also extremely fertile in returning effective interpretative frameworks of the dynamics of innovation are those 'guided by design'.

Also of great interest in a learning programme aimed at producing innovative acts through design are *historical studies*. This is for various reasons.

The time scale, the area of observation and also the method used by history to look at innovative phenomena, poses the problem of the direct link between intention and action, overturning the typical assumptions of a theory, the player and the action, where every single act finds an explanation and a paternity which are easy to identify. Historical narration on the other hand accepts the undesired consequences, mitigating the links between decision and action or making them products of complex plots that cannot be reconstructed. This arrangement of historical studies has also had a significant influence on the sociological analysis of innovation processes, opening the minute study on the work of the conceiver on the analysis of the great evolutionary currents of technology and sociology which build up the reference frameworks and imaginary of an age (Flichy 1996).

Of particular relevance in this sense is the historiographical arrangement developed in France, starting with the work of Braudel (1975), not a story of events and personalities, but a story of mentalities, legends, rituals, beliefs and values that restore to the context its exceptionality, in the sense of being a 'unique situation' capable of being an explicative reference framework of equally unique actions, which are interesting because they are unique.

In other words, historical studies result are important in forming a tendency in students and an ability to read the big contexts in which thoughts, styles, actions and events are placed as a starting point for re-elaborating new imaginaries and new values through design.

We have already looked at the subject of the imaginary on several occasions. This is a recurring subject in literature with the same superficial approach as that found with regard to the term creativity.

However, I would like to open a parenthesis on the importance of this term within a learning programme, because I think it can make interesting contributions to the practices of *advance design* and, in general, to those passable actions of innovation and transformation that can introduce changes with a vast technological, social, cultural and useful impact.

Once again in this case we find ourselves facing an extensive repertory of studies that have created the definition and phenomenologies (psychoanalysis, psychology, social psychology, sociology, anthropology, etc.).

Here are just one or two brief considerations.

According to many authors interested in innovation processes, the subject of the imaginary is an important reference for analysing the inventive phase, where utopias, dreams, myths, imaginary projections and cognitive materials—often of visual nature—related to the culture and sensitivity of inventors and designers form an important matrix of reference in terms of technological innovation and formal innovation. But the subject of the imaginary become substantial when, during the dissemination of an innovation, *the individual imaginary becomes collective* and forms a sort of *common sense* that makes change acceptable.

So we have a circular process in which the imaginary of the inventor is nourished by common sense that transforms it but, in order to produce change—and for the individual imaginary to become collective heritage, it is necessary to arouse interest, resources of numerous players who participate in the elaboration of a social imaginary, through *forms of discourse* that expand acceptance by the specific world of that innovation to contiguous worlds (Legrenzi 2005).

It is interesting to take from the sphere of social sciences, that combination of reflections known as *theory of social representations* that operate on the way in which knowledge is represented in a group, shared by its members until it becomes a form of common sense.

The theory of social representations was developed mainly in France and was presented in its complete form for the first time by Serge Moscovici nel in 1961

(Moscovici 2005). In his theoretic research, Moscovici looks at the way in which knowledge is represented in a given social and shared context, investigating the processes through which individuals elaborate images that help make the non-familiar, familiar.

Obviously I'm not going to spend time here supplying more precise references to the subject. I would just like to add that, in the social exchange that leads to the sharing of new concepts, a relevant contribution is made by those forms of exchange that use allegoric figures, allusions, metaphors, verbal and visual materials that have a strong symbolic content that builds a sort of bridge, linking old and new (Benoist 1976).

This brings us back to the initial assumption: even the most advanced forms of innovation—and especially the forms of innovations that presume a string detachment from consolidated contexts—always need to 'bargain' with the context they fall within. Particularly, with the social context it is called to accept and to attribute sense to change.

References

Amietta, P.L.: I luoghi dell'apprendimento. Metodi, strumenti e casi di eccellenza delle nuove formazioni. Franco Angeli, Milano, 2000

Bara, B.G.: Il metodo della scienza cognitiva. Bollati Boringhieri Editore, Torino (1999)

Belo Reyes, P.: Projecting from the exteriority of the project: a contribution to the teaching processes in project design. In: International Forum of Design as a Process, Torino, 2011

Benoist, L.: Segni simboli e miti. Garzanti, Milano (1976)

Braudel, F.: Civiltà materiale, economia e capitalismo. Einaudi, Torino (1975)

Ceriani, A.: La simulazione nei processi formativi.Una metodologia per un pensiero creativo progettuale. Franco Angeli, Milano (1996)

Ceruti, M.: Evoluzione senza fondamenti. Laterza, Roma-Bari (1995)

Davenport, T.H., Prusak, L.: Working Knowledge: How Organizations Manage What They Know. Harvard Business School Pr, Boston (1997)

David, P.: Clio and the economics of QWERTY. Am. Econ. Rev. **25**(2), 332–337 (1985)

De Fusco, R.: Il design che prima non c'era. Franco Angeli, Milano (2008)

Emery, F.E. (ed.): Systems Thinking. Penguin Books, Harmond- Sworth (1969)

Feibleman, J.: Pure science, applied science, technology, engineering: an attempt at definitions. Technol. Cult. **2**(4), 305–351 (1961)

Flichy, P.: L'innovazione tecnologica. Feltrinelli, Milano (1996)

Frascara, J.: Design Research as an action-oriented interdisciplinary activity. In: Pizzocaro, S., et al. (eds.) Design Plus Research, pp. 13–15. Atti di Convegno, Milano (2000)

Georghiou, L., et al.: Post innovation performance. Technological development and competition. MacMillan, London (1986)

Johnson-Laird, P.N.: The Computer and the Mind. An Introduction to Cognitive Science. William Collins, London (1988)

Kuhn, T.S.: The Structure of Scientific Revolutions. University of Chicago, Chicago (1962)

Legrenzi P.: Creatività e innovazione. il Mulino, Bologna (2005)

Maldonado, T.: Scienza e progettazione. Avanguardia e razionalità, pp. 179–199. Einaudi, Turin (1974)

Maldonado, T.: Il futuro della modernità. Feltrinelli, Milano (1987)

Morin, E.: La testa ben fatta. Raffaello Cortina Editore, Milano (1999)

Moscovici S.: Le rappresentazioni sociali. il Mulino, Bologna (2005)
Nelson, R.R., Winter, S.G.: An evolutionary theory of economic change. The Belknap Press, Cambridge (1982)
Nonaka, H., Takeuchi, H.: The Knowledge Creating Company. Oxford University Press, New York (1997)
Oliverio, A.: Memoria e oblio. Rubbettino, Catanzaro (2003)
Penati A.: Mappe dell'innovazione. Il cambiamento tra tecnica, economia e società. Etas, Milano (1999)
Penrose, E.T.: The Theory of the Growth of the Firm. Blackwell, Oxford (1959)
Petroni, G.: Leadership e tecnologia. La matrice organizzativa della grandi innovazioni industriali. Franco Angeli, Milano (2000)
Pizzocaro, S.: Utente/User. In: Pizzocaro, S., Figiani, M. (eds.) Argomenti di ergonomia, pp. 169–186. Franco Angeli, Milan (2009)
Rogers, E.: Diffusion of Innovation. The Free Press, New York (1983)
Sasso A.: Introduzione. In: Sasso A, Toselli S. (eds.) La scuola nella società della conoscenza. Formazione, tecnologie, informazione, modelli di vita. pp. 1–31. Bruno Mondadori, Milano 1999
Schumpeter, J.A.: Business Cycles: A Theoretical, Historical, and Statistical Analysis of the Capitalist Process. McGraw-Hill, New York (1939)
Simon, H.A.: The New Science of Management Decision. Harper and Row, New York (1960)
Simon, H.: The Sciences of the Artificial. MIT Press, Cambridge (1969)
Skolimowski, H.: The structure of thinking in technology. Technol. Cult **7**(2), 371–383 (1966)
Usher, A.P.: A History of Mechanical Inventions. Beacon Press, Boston (1954)

Part II
Phenomenology of *AdvanceDesign*

Chapter 6
Reading *AdvanceDesign* Practices

Elena Formia and Danila Zindato

> *Given that external reality is a fiction, the writer's role is almost superfluous. He does not need to invent the fiction because it is already there.*
>
> James Graham Ballard

6.1 For a Qualitative Study of *AdvanceDesign*

The general aim of this book, and of the pertinent research, is to explore the wideness of the research field in *AdvanceDesign*. The use of the word exploration means that the approach has not been fully structured, directed or programmed; the research has been carried out as a sort of constant, choral brainstorming. What is *AdvanceDesign*? How is it done? Who does it? Why? Where? Since when? From this viewpoint, the five essays that follow examine possible scenarios of application of *AdvanceDesign*, documented through the presentation of case histories. This methodology, which is particularly used in the sociological and economic fields, adapts effectively to the research *through* design[1] especially during an embryonic

[1] The terms 'research on design', 'research through design' and 'research for design' were proposed by Frayling (1993) and then discussed by many authors. However, this interpretation seems very appropriate: 'So, design research is beginning to be interpreted from a phenomenological standpoint, observing the reality of design to extract general rules and principles that are, however, in constant evolution with changing viewpoint and context […] From an operational point of view, it is not only or primarily the academic community that explicitly interprets this approach, but actors in the professional world that, increasingly, not only design but also acquire and develop knowledge about and for design' (Bertola and Maffei 2008, p. 16).

E. Formia (✉)
Alma Mater Studiorum, Università di Bologna - Dipartimento di Architettura,
Viale del Risorgimento 2, 40136 Bologna, Italy
e-mail: elena.formia@unibo.it

D. Zindato
Dipartimento di Design, Politecnico di Milano, Via Durando 38A, 20158 Milano, Italy
e-mail: danila.zindato@gmail.com

M. Celi (ed.), *Advanced Design Cultures*, DOI 10.1007/978-3-319-08602-6_6

stage of the investigation, when the intention is to generate hypothesis and reserve in-depth analysis later on.

The overall process included a first study phase during which, through a continuous socialization of partial results by the research group, it was made an attempt to understand the problem and build up an exploratory picture that formed the common filter for the collection of information. The second part of the book offers, on the other end, a description of the problem inside different contexts of application of *AdvanceDesign*'s methods, practices and instruments. Each author involved was free to identify the main features of *AdvanceDesign* based upon his/her personal experience, choosing two or more case histories capable of portraying the hypothesis discussed. Nicola Crea (Chap. 7) presents the profile of the advanced designer in the automotive sector introducing the structures set up by Mercedes-Benz and Fiat; Raffaella Mangiarotti (Chap. 8) focuses on the role played by *AdvanceDesign* in the process of meaning generation, narrating two *scenario building* models implemented by Philips and Whirlpool Europe; Maurizio Rossi (Chap. 9) proposes an experience-based research method that envisages a new approach in which lighting design is considered strictly in relation to the human factor; *AdvanceDesign* plays a fundamental role in the changes taking place in the organization of companies and their links with the territory in the essay by Stefania Palmieri (Chap. 10); finally Marinella Ferrara (Chap. 11) presents how design and techno-science are already able to communicate through *Advanced Material Design*.

This spontaneous approach led to the gathering of a heterogeneous corpus of information and to the exploration of various dynamics between research and design. This is mainly due to the different backgrounds of the authors involved. Pure researchers analyzed some experiences dedicated to specific areas of study in the design field (i.e. advanced lighting design, advanced transportation design, advanced material design, etc.). Others, who also have a huge professional experience as designers and have carried out specific projects in the field of *AdvanceDesign*, presented research matured in productive sectors with a pioneering tendency towards the application of this approach.

In the process of continuum debate and of construction of contents, the group also opened its research developing a network of relationships between different kind of subjects (professionals, companies, research centres, etc.). They were involved in specific projects or simply chosen as privileged witnesses of *AdvanceDesign* approaches. This helped in animating the discussion about the different ways in which *AdvanceDesign* may support the development of instruments and practices aimed at managing value (Celaschi et al. 2011; Celi 2012; Celi et al. 2013), about how it may be connected to the ethic implication of the design activity, or, finally, about why it may be related to the visionary dimension of the discipline (Deserti 2009). For example, a strong collaboration was established with the Milan based consultancy firm Design Innovation. The studio, founded and directed by Carmelo Di Bartolo, has a thirty-plus years experience in different sectors, during which it has developed a method of knowledge cross-fertilization based on an interdisciplinary working group (designers, engineers, economists and experts in specific fields according to specific project's demands). This experience led to the exchange of resources with other geographical and cultural contexts, such as the Advanced Design Center hosted at the Tec (Tecnológico de Monterrey) Campus Guadalajara (Mexico), as a place to give local and national industry a space with the necessary facilities for developing new products and services; or the Department of Sociology and Social Research of the University of Trento, directed by professor Roberto Poli. New research interests have consequently emerged: for example, *AdvanceDesign* and the Future Studies' field can establish areas of comparison between methods, processes, practices, such as anticipatory thinking protocols, causal layered analysis (CLA), environmental scanning, future history, monitoring, eco-history, back-view mirror analysis, cross-impact analysis, futures biographies, etc.

The result of this process is that the subject of the study has not been analyzed only from one point of view, but from a variety of angles, which produced a multifaceted picture of the phenomenon. However, it is possible to establish a relation between the theoretical reflection and the applied research by classifying some variables and finding similar characteristics and results.

6.2 A Synthesis of Case Histories Through a Map

This introductory chapter provides a guide to the material within the second part of the book and, at the same time, creates a *trait d'union* between the two sections of the volume. This is achieved by focusing on the twelve case histories outlined in the phenomenological analysis by the five authors, visualizing them on the map of *AdvanceDesign*'s dimensions (Chap. 1). The analysis starts with a brief summary of each case followed by an explanation of its correspondence with the directions outlined in the general map.

1. Mercedes Advanced Design Studios: Mercedes-Benz is one of the automotive company that historically recognized the *AdvanceDesign* process as integrative part of the new product development (NPD) process giving birth to three consultancy studios worldwide: Carlsbad, Yokohama and Como. The purpose was to trigger the processes of cross-fertilization between the mother company and the local cultures in order to strengthen the brand identity in a perspective of increasing globalization.

Positioning: Time Factor. This case presents the typical feature and capability of the transportation design sector to create systems for identifying trends and future markets from different points of observation.

2. Fiat Advanced Design: The Italian automotive company opened an *AdvanceDesign* centre to develop new products and strategic plans with a broad temporal horizon. The aim is to adjust the various phases of the design process and have a set of constant feedback both from outside and inside the company. The ADUS (Advanced Design University Stage), purpose-built for training young professionals, was established inside the centre to analyze design problems and create innovative contributions for the company.

Positioning: Time Factor. This case has strong relations with the time factor insomuch as the *AdvanceDesign* centre identifies possible future scenarios related to the car through a planned and integrated process.

3. Philips Vision of Future: The project, aimed at the definition of future scenarios, was oriented to the spheres of domesticity, person, public and mobility, with a time horizon of 15/20 years. The research was nourished by interdisciplinary skills that mixed also socio-cultural and technological areas. The outputs were subsequently filtered and validated by international opinion leaders. Finally, experts of interactive storytelling integrated the final results, embedded in several prototypes, in order to obtain a clearer feedback from the consumers.

Positioning: Time Factor and Without Market. Even though oriented to the envisioning of future environments in different fields, the project is a self-assigned research and does not respond to any market demand.

4. Whirlpool Europe Project F: The Whirlpool Global Consumer Design (GCD) is dedicated to research at different levels, from the identification of new opportunities to strategic plans and new products development. The Project F had been conducted across three areas and its goal was to investigate textile care for innovating the washing machines. The process was divided into phases (workshops, prototyping and communication) and was developed by an integrated group formed by three worldwide consultancies (Germany, Italy, the United States) and six in-house designers.

Positioning: Time Factor and Business to Business. The project is future-oriented, but the results are semi-finished products created by professionals for an internal use and for the 'mission statement' of the company. The advance design experience, in this case, can be also considered as a process of renewal and training.

5. Fisiolux: The research activity was carried out by the Light Laboratory (lighting, photometry, colour and perception) of the Politecnico di Milano on behalf of the Italian lighting company Artemide SpA. It was characterized by two features: the interdisciplinary nature of the project team and the transfer of skills and knowledge from one sector to another, in order to identify and apply the concept of welfare systems in residential lighting. The research led to the development of the product line My White Light.

Positioning: Spatial and Sectorial Factor. The project is focused on interdisciplinarity and on the exchange of expertise among the sectors involved (psychology, medicine, design).

6. Food Lighting: The project, committed by Castaldi Illuminazione SpA, was performed by the Light Laboratory of the Politecnico di Milano with the aim of identifying the best lighting solutions in food stores. The research was carried out analyzing the different areas involved in the relationship between light, food and human being: the technological aspects played the same role as the psychological and perceptual one. The project involved various research fields and numerous user groups for different experimental trials.

Positioning: Spatial and Sectorial Factor. As the previous one, this case is the result of a cross-fertilization process between user experience, perception, colour and design for creating innovative lighting appliances.

7. Campus Innovazione Automotive: Since 2008, the IAM consortium (Innovation Automotive and Mechanical engineering) has worked in the Sangro Aventino district (centre of Italy) to maintain the know-how and research activities within the territory. Among the initiatives, there is the creation of a campus in which operational and methodological tools are shared. The aim is to consolidate the entrepreneurial sector of the automotive industry and strengthen the development of business networks.

Positioning: Business to Business. The network aims at bringing a know-how exchange among enterprises working in the same area but with different core business, in order to optimize the processes and to create new hybrid paths of innovation.

8. Kilometro Rosso: The scientific and technological park launched by Brembo is a perfect model of multidisciplinary *Research & Developement area* oriented to the consolidation of a knowledge, innovation and technology district. It is based on the idea that, for producing a real innovation, it is necessary to create integration among different sectors and realities. For this reason the centre fosters multidisciplinary projects that give rise to cross-fertilization processes and, more generally, to *AdvanceDesign*.

Positioning: Business to Business and Spatial and Sectorial Factor. The network has a common technology platform, but it is made by different business models that work at different levels. The technology park also furnishes spaces as facilitators of innovation processes.

9. GEIE ECSA: The ECSA (European Centre for Space Applications) is a scientific network comprising universities, SMEs and research centres with the aim of creating an exchange system of collaboration. The common objectives are the exchange and transfer of skills and the achievement of a proper dimension to compete in strongly internationalized markets.

Positioning: Business to Business and Spatial and Sectorial Factor. Compared to the previous one, the network presents enlarged fields of action, even if the aim is always to produce useful results for the companies themselves and not to create products/services to the end customer (B2C).

10. Materialecology: Since 2006 Blacks Oxamn has built, in Cambridge, the research centre Materialecology. Its interdisciplinary aim is to study and model natural structures and behaviours in order to extract geometries, solutions and systems and carry them in the industrial sector, combining a biomimetic vision to design.

Positioning: Spatial and Sectorial Factor and Without Market. The transfer of knowledge and models from one sector to another does not take place for a precise need of the market or request of the client, but in an experimental context.

11. Material Beliefs: The group based at the Department of Design of the Goldsmith University of London is focused on the results achieved by research in the biomedical and cybernetics fields and their possible applications in every day life. The contamination of knowledge among different fields of study, from engineering to social sciences, leads to the design of prototypes which embed these parallel outcomes into something tangible. These prototypes are exhibited, transforming an emerging laboratory research into a platform that encourages a debate about the relationships between science and society.

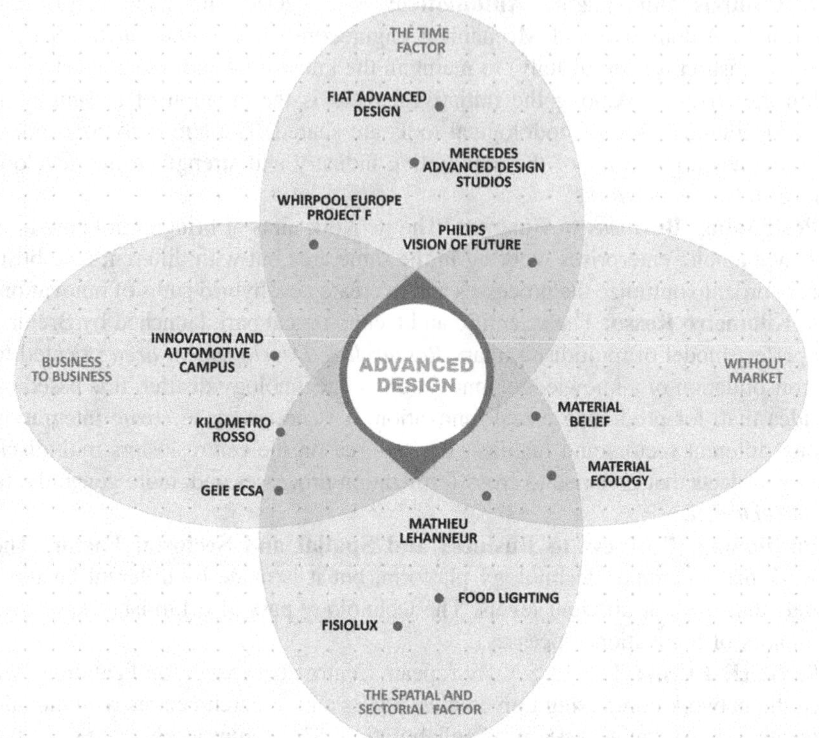

Fig. 6.1 The twelve case histories visualized in the map of Advanced Design's directions as interpreted by the research group

Positioning: Spatial and Sectorial Factor and Without Market. The transfer of technologies and solutions from a field to another field auto-generates innovation in processes and products, even without a specific market demand.

12. Mathieu Lehanneur: This case concerns a designer and not a research group, a company or a network. He has been selected for the ability to create overlapping zones and knowledge contaminations between different sectors, such as the realization of a set of objects obtained mixing the results of NASA studies, the world of medicine and the design field.

Positioning: Spatial and Sectorial Factor and Without Market. The transfer of knowledge from one field to another is the basis of the researches carried out by the French professional.

The positing of case histories in the map should not be intended as a simplification of the research outlined by the authors of the second part of the book. The map underlines the descriptive dimension of the phenomenon as intended by the research group. However, there is no rigid compartmentalisation, just a possible guide to reading. This is proven by overlapping and interference that characterize the experiences described and visualized in the map (Fig. 6.1).

The purpose of the actual qualitative investigation is to isolate those weak signals that can be used as reading keys to interpret the future. A sort of 'present-day science fiction' with a Ballardian flavour: futuristic tendencies transformed into current elements of our experience.

References

Bertola, P., Maffei, S.: Design Research Maps. Prospettiva della ricerca universitaria in design in Italia 2003–2007. Poli.Design, Milano (2008)

Celaschi, F., Celi, M., Mata García, L.: The extended value of design: an advanced design perspective. Des. Manag. J. **6**(1), 6–15 (2011)

Celi, M.: Ricerca, visualizzazione e scenari per l'innovazione di processi e prodotti. In: Indesit & Design Innovation (eds.) Materials Driven Design, pp. 24–43. Fausto Lupetti–Dodici edizioni, Milano (2012)

Celi, M., Iñiguez Flores, R., Mata García, L.: Design as value catalyst for SMEs in emerging contexts: the case of Guadalajara, Mexico. J. Des. Strateg. **6**(Spring), 45–55 (2013)

Deserti, A.: Design, craftsmanship, art: Liaisons Dangereuses? In: Atar, A., et al. IMECE 2009 Fine Arts & Design Symposium with International Participation. Anadolu Üniversitesi Güzel Sanatlar Fakültesi, Eskisehir, pp. 7–14 (2009)

Frayling, C.: (1993/1994) Research in art and design. R. Coll. Art Res. Pap. 1(1): 1–5

References

Chapter 7
Advanced Transportation Design

Nicola Crea

From how to cut paper to where did I put the scissor.
Alain Fletcher

The automobile is expected to evolve with respect to how it is conceived today. There will be new ways of moving, new forms of ownership and management, new morphological configurations, intended for more specific uses. Especially from the construction point of view, to reduce implied costs of logistics and warehouse, the automobile will be composed of interchangeable parts belonging to a modular construction system. Easy assembly and disassembly will allow easy repairs, recycling of components, restyling, personalization and aesthetic updates. These are some of the parameters with which the car design must confront in order to update its production. But the themes of innovation can be much broader. The car manufacturers to keep current and competitive in the global market have entrusted the task to specific facilities to handle those issues that are typical of the *AdvanceDesign* and that do not bring to an immediate economic return. The reasons why the first companies that have begun to use the *AdvanceDesign* procedures have been just the automakers may be different. Of course, for the development of a new vehicle it takes about 3 years and several millions of euros. The investment cost for setting up a production line is therefore a very high-risk business. The desire to reduce this risk to be taken, jointly with market analysis previews, requires the establishment of operating procedures that protect their investment. To reduce the financial risk, you must anticipate what will be the cultural context and the market in which vehicles will sell at launch, so as to ensure a fair approximation of commercial results. In fact, to test the response of potential users, car companies, before embarking on the development of new models, propose concept cars at auto shows.

The design of transportation means, in addition to deal with technical and formal innovation of the product, faces collaterally some problems that demonstrate the complexity of the project. The factors that complicate and affect the definition of a new product are varied. For example, compliance with laws and regulations binding

N. Crea (✉)
Dipartimento di Design, Politecnico di Milano, Via Durando 38A, 20158 Milano, Italy
e-mail: nicola.crea@polimi.it

M. Celi (ed.), *Advanced Design Cultures*, DOI 10.1007/978-3-319-08602-6_7

on the projects, set homologation schemes that affect the configuration of the vehicle. These, in fact, fall into categories of performance and dimensional characteristics, which are somehow predetermined. The implementation of many solutions is bound by existing patents. Environmental degradation requires vehicles as recyclable as possible and less polluting. There are numerous rules regarding consumption, recycling means at end use and the morphologies allowed by manufacturing processes. These constraints affect the choices about technologies, processes and the use of specific materials. Furthermore, control of costs weighs heavily on the project. Increasingly selective and competitive, the current global market demands high-quality products at lower costs. The economy and the market, therefore, are essential factors in defining content and design features. The control and containment of investment and production costs, on the one hand and the study of production synergies and maximum use of 'carry over' components on the other hand, significantly affect the final configuration of the product. The list of constraints in the formal definition of a vehicle could be even larger if we take into consideration the respect for the image of the product, the established values of the brand and product positioning related to the competition. These considerations bring consequently a notable lack of freedom in design.

The clear and progressive increase in complexity of new projects development and the relative increase in required professional skills, result in a greater participation of designers in the processes of 'product development'. This means, on an operational point of view, the integration of designers works with structures dedicated to engineering and industrialization. Consequently, in these cases, the designer moves away from its prerogatives of 'visionary', inventor and producer of the new (Del Bono 1996), to be involved especially in further downstream problems, part of the execution phases of the development process. To achieve constant innovation, though, businesses have to face a highly competitive and evolving market. It requires, however, constant experimentation, on a practical and conceptual side, and research of new joints to refresh the salability of the product in respect to the contingent needs of the market. These indications have produced opposed guidelines to the role of design inside industry. It has led progressively to specific roles of specialization according to business needs. Besides different definitions that have been given, we can say that design, if you intend it simply as a creative process, a summary of given parameters can be expressed in many different forms. It can range from graphic design to fashion, from packaging to the design of boats. Today, for example, we clearly recognize independent and specialized areas such as product-design, interior-design, transportation-design, graphic-design, fashion design, etc. Depending on the subjects treated, which are substantially different in respect to skills, work procedures and commercial sectors of reference, we can recognize very specialized areas such as, for example, yacht-design, car design, exhibit-design, packaging-design, etc. The professional specialization, in extreme cases as strategic design or design management, even carries the paradox to set aside the actual design of products, but to perform a function of planning, strategic vision, coordination and control of processes within the company.

7.1 Design for Product Development and Advanced Design

In the specific field of car design, we can delineate two new professional profiles with different operating characteristics: the designer who is dedicated primarily to product development and the specialist of Advanced Design. Both represent an evolution of the role usually taken by car designers. It arises by a technical-practical work configuration, for the first one and by theoretical-conceptual and more strategic contents, for the second one, respectively. Both roles are emerging, identified to meet their specific needs. The role of the designer specialized in product development arises from the need of organizations to reduce the 'time-to-market.' This goal can be achieved by integrating stages of design and style definition with product development activities. Thanks to this opportunity and an ongoing confrontation, it is possible to avoid the reiteration of the creative process, by immediate verification of the adequacy of the choices made. This way, creative activity and design work of the designer develops in parallel with the phases of engineering and industrialization, from the definition of the concept until starting of production. The designer, engaged into solving problems in the product development process, must anyhow maintain the respect for other requirements. These can be technical, aesthetic, cultural and cost needs, as well as market positioning, product identity and brand image. Who is involved in product development, for obvious reasons, is to neglect some of his prerogatives of creative and innovative person. Necessarily he will end up finally in solving problems aimed at the industrialization of the product. On the other hand, the most speculative and experimental research activity must be operated by another professional. Generally, this is expressed by the Advanced Designer. Beside the operative designer, as figure integrated into the process of product development, which is an expert in human relationships and compromises, industrial technologic processes, in the various components and materials to be used and mainly committed to the technical aspects and embodiments, it is felt the need to seek the contribution of another designer with more freedom, less affected by project ties, but equally competent. Therefore, facilities for conducting studies of Advanced Design have been established. Most car manufacturers have committed human and financial resources to the Advanced Design, setting up a specific professional role, just to ensure that creative and purposeful function that threatened to overshadow. We can imagine a figure of designer projected in the future, able to see farther: a 'concrete' visionary, able to invent new product concepts potentially feasible, to imagine new functions and new methods of use for existing products, to provide new scenarios, to 'create' without being conditioned by immediate expectations of the market (Del Bono 1996). This, briefly, describes the main features that distinguish the activities of specialist designers of Advanced Design.

In the automobile sector, in a scenario strongly influenced by constraints, the Advanced Transportation Design provides the desired opportunity to design regardless the number of constraints mentioned above. Autonomy from temporal, economic and market requirements allows a much wider room for testing and evasion from the predictable logic of development, and closely tied to immediate economic and technical opportunities. The Advanced Transportation Design must offer a service tailored to innovation, that allows the application of scientific discoveries and technological developments, the use of new materials, a new way of conceiving products beyond the regulatory and commercial constraints. The results produced may not be immediately usable, but they will be the essential propulsion to innovation, a wealth of knowledge and a wealth of ideas from which to draw at the appropriate time in order to obtain an impact on product sales.

7.2 Technology Transfer

The Advanced Design, in pursuit of its objectives of innovation, requires the use of sophisticated technology, in both the product development process and the production process. Some of these technologies can be already acquired skills (part of the assets of business or in form of professional knowledge of the designer), or may have already spread to other disciplines or product and then easily be found. From a methodological point of view, it can be of great help to carry out a survey of the technologies used in various other disciplines such as, Medicine, Chemistry, or, to quote a particularly advanced field on a technological standpoint, the aerospace engineering. It would be convenient to carry out a survey on what could be transferred in its field of application, both in technical and economic terms, even if only by analogy of operation. Tools and techniques with a high level of diffusion and wide application in a given commercial sector, once re-used in another area may favor the transfer of knowledge and, consequently, induce innovation. The immediate benefit deriving is that you can rely on knowledge and equipment that has been developed and already tested elsewhere to satisfy your needs (Arielli 2003). From the classic example of the invention of penicillin on, the whole history of scientific progress is littered with incidents or deliberate technology transfer. But much more can be done. The use of probes equipped with a camera for investigations into the human body used in medicine can be of great help in the investigation of inaccessible areas of buildings to identify breaks or leaks, avoiding 'to break' walls or wall coverings. Another example is that of Rovema, a German company leader in the construction of machinery for packaging of light products. At a glance, the problem faced was to find an effective way to quickly pack potato chips and keep them whole. The problem of inserting the chips inside the package without breaking them is similar to the problem of landing a spacecraft on a planet with no damages. Working on this similarity and using the skills developed by the industry that had been working on ESA space projects, Rovema has found a way to conceive a more efficient machine for the packaging of delicate food.

7.3 Interdisciplinarity

Technology transfer and long-term forecasts are not sufficient conditions to define products with *AdvanceDesign* criteria. In fact, this does not necessarily mean that they also possess the unique features, functional, structural, or technological, which constitute true innovation. To achieve these goals, a broad and diversified expertise is required. The innovation research assumes knowledge and relationships among disciplines very distant from each other (Chiapponi 1999). This is because the structures that carry out research aimed at innovation typically involve professionals and specialists from different disciplines. They consist of mixed teamworks, where engineers, chemists, physicians, ergonomists, and whoever it would be necessary to ask for their specific skills, work together. We must emphasize that, in the case of the automobile, the component of engineering is conspicuous. The relationship between design and engineering, in this case, is very tight. The dialectic of innovation is founded precisely on the inclination of the engineer towards the new and the depth of technical expertise of the designer. A remarkable example is the design of the old Citroen DS. The car, at the time of its presentation (1955), was very successful for the quality and quantity of innovation introduced in the project. As many enthusiasts already know, the car was equipped with several novelties for the automotive sector. Among these are worth mentioning: the plastic material roof, aluminium front hood, a flat floor, an adjustable geometry suspension system, all body panels bolted to the structure, always in black colour, the doors without window frames, turning headlights, back turn signals on the roof, a special water pump for windshield washer, the possibility of changing the wheel without the use of jacks and even the chance to travel with only three wheels removing one of the rear wheels, not bad for a car designed 60 years ago. In this fortunate case, the designer Flaminio Bertoni integrated the designer's contribution to an interdisciplinary project team that produced perhaps the car with the highest number of innovations that has ever been produced.

7.4 The Role of University

As we have seen, the *AdvanceDesign* research, having innovation as its major goal, it is expressed by its interdisciplinary nature. The teams involved are usually supplemented by the contribution of external consultants, or specialists coming from leading centres of competence. The collaboration of faculty and university departments, with their own research units in their various specializations, can provide valuable support. This practice is common in most industrialized countries, like Germany and the United States, where the relationship between university research centres and industry plays a normal role in the innovation process. Firms are routinely engaged in solving current problems and do not have the resources to grants for research and experimentation. A permanent connection between research and industrial sector combines the needs of innovation with the need for research to find application opportunities. Today, in the university, you look with particular favour at relationships that can be established with businesses and industrial districts. The *AdvanceDesign* is a matter of common interest, which should be the meeting place between academia and business.

7.4.1 Experimentation

The *AdvanceDesign* today is essentially a specialization of industrial design, facing long-term objectives and implemented with experimental procedures. It is not to design for the present time. They foreshadow possible future scenarios, which in some cases differ considerably from actual conditions. Design issues are addressed in a radical way, unhooked from the binding effect of commercial and, possibly, economic nature. The *AdvanceDesign* expresses a speculative and pure research activity, untainted from the many factors considered restrictive elements of innovation and as background elements for the sheer resolution of technical problems. Apart from the positive values, implicit in the broad connotation of the term that identifies it, the *AdvanceDesign* definition expresses a very in-depth specialist expertise. This happens not only for the theoretical and conceptual phases, but also for the technical aspect and the application. Generally, projects developed by the *AdvanceDesign* criteria, end up in the construction of prototypes and demonstration models destined to be displayed in trade fairs and exhibitions. Content can be the experimental use of new materials, the use of new technologies or the application of new functional concept. In essence, they are research projects not intended to be produced in the immediate. Sometimes they have advertising purposes or they serve as survey of a potential market. In many cases, after an appropriate process of industrialization, the proposed innovative concepts have even found commercial opportunities (Crea 2002). By way of example, it is worth to mention the project of windsurf board, born without the support of a real market need, and its following developments (paragliding, kite-surfing, etc.), or the design of the Smart, born in the wake of 'Swatch phenomenon'. It has been responsible for responding to an objective need for mobility and parking with the development of a specific product.

7.4.2 Cases of Mercedes-Benz and Fiat Auto

The specific contours of the *AdvanceDesign* as a discipline are to be found in the broader theoretical framework related to industrial design whose definition is a object of a still very open discussion. According to Maurizio Vitta, and his hypothesis of a 'mobile' theory of design (M. Vitta 1996), we could also say that the *AdvanceDesign* will never have a definitive identity. Compared to industrial design, meant simply as the work related to shape definition of an object industrially produced, *AdvanceDesign* has a more complex connotation. We were able to recognize some features easily identifiable as the foreshadowing of long-term scenarios and the definition of strategic lines of development within the company. However, there is still a considerable margin for a further definition, because each one understands this specialized activity to suit their needs. Cases like Mercedes-Benz and Fiat are two significant examples. Even starting from similar assumptions, the two automakers have made, for many years, facilities for implementation of Advanced Design, in a very different setting from each other. For Fiat to see further, it is intended to involve young people in the teams for advanced research. For Mercedes, the objective was to study different cultures and the survey of new markets. Fiat wanted to integrate its advanced research with the opportunity to select new talents that can envision future scenarios. Fiat, with the help of Design Innovation of Carmelo Di Bartolo, gradually reached the goal of consolidating a structure used for the creation of innovative concepts. Mercedes, on the other hand, has established a network of studies in three different continents in order to govern a world observatory on the cultures and trends in different geographic areas. For Mercedes, to explore these possibilities of tastes, preferences and inclinations of distant markets are justified by the breadth of its commercial destination. It demonstrates also with how much concern of vision and foresight are designed future vehicles. The charts that follow are showing the characteristics and content of the corporate structures that Fiat Group Automobiles and Mercedes-Benz allocated to the Advanced Design.

Case 1: Mercedes Advanced Design Studios

The case of the Mercedes-Benz is emblematic. It has integrated advanced design as a standard process in the development of new products, and it has established an international network of design studios. These structures finalized to the practice of advanced design provide a global overview of the development of Mercedes-Benz. Placed in strategic places, like Italy, Japan, the United States and Germany, of course, constitute a privileged 'observatory' of the main world car markets. New concepts arise regularly from Advanced Design. Possible future configurations are proposed to the critic of the public in form of show cars. In case of positive response, then, the projects are developed to be brought to production. This is the case for example of 'class A' and 'SLK' which turned into successful products.
In 1990, as a part of the international orientation of its design strategy, Mercedes-Benz opened its first foreign advanced design studio in the United States, Irvine, California. Since then, the importance of the California studio has grown to such an extent that in a short time room available was no longer sufficient. Mercedes-Benz has decided than to move in July 2008 to a new building with a much larger floor area, in Carlsbad, 80 miles south of Los Angeles. More than 10,000 km away from the main office in Sindelfingen, Carlsbad, with about 25 employees from different cultural backgrounds, is an

ideal working environment for creative expression and for the development of after-tomorrow Mercedes-Benz design. The centre focus on the development of complete vehicles and it is equipped with tools that allow the possibility of building 1:1 scale models.

As long as with the American studio, there are also other locations which can be considered as creative islands not affected by production requirements. These privileged observation points allow designers to perceive the local environment and cultural context, in order to evaluate and interpret it freely. The inspiration provided is an important lever for development of Mercedes-Benz design, which is not oriented simply to the domestic market, but it expresses a global vocation. The aim is to develop new ideas, originate from different cultures, in other areas of the world, which can enrich Mercedes-Benz identity. In fact, one of the main difficulties of Mercedes-Benz design philosophy is its ability to remain faithful to the stylistic tradition of the company, while proposing new ideas. These several international influences, however desirable, must anyhow combine with the already established brand values.

In 1992, Olivier Boulay was entrusted to the opening of an advanced design centre for Mercedes-Benz in Japan. It was a part of the policy of 'globalization' adopted by the German manufacturer to adequately face future challenges, especially in this part of the world. Japan has been chosen by Mercedes for the creation of an advanced design centre also because of the local technical environment and the presence of infrastructure. Japan, with nine automobile manufacturers, three motorcycles manufacturers, various other commercial and industrial vehicles factories as well as a large number of suppliers, still represents a most promising location for the study of future technological and stylistic innovations in the automobile industry. The frequency of renewal of Japanese products, due to the strong competition, is really high. It leads local markets to follow new trends with great speed and flexibility, and to abandon them just as quickly. These reasons have caused in time a large development and production capability and a very articulated demand.

The headquarters of Yokohama is a platform for studying the Japanese market, but also an observatory on the entire East Asian market, currently undergoing rapid development (two-thirds of world population). Japan is the dominant part, having production widespread also in Europe and in the United States, but the markets of China and Korea are growing very quickly as well.

To establish an international team, in the Japanese studio were hired 15 designers, modellers and highly qualified technicians mostly Japanese, of course, but also Chinese and Vietnamese. To Western culture dealing with different Eastern cultures, languages and lifestyles represent a real challenge. The most delicate and yet most fascinating goal is to see these different integrated cultures cooperate constructively and creatively with other similar structures, in California, Germany and Italy. The contribution of ideas provided by the Tokyo design team to Mercedes Design is generally high, but the most recognized and most important result has been the development of the new Maybach brand, born for the connotation of automobiles of extreme luxury.

Opened in February 1998, the Advanced Design Studio in Como has specialized in Interior Design. The choice of Como as third location, after Carlsbad and Yokohama for Mercedes-Benz Advanced Design has also been dictated by considerations of a strategic nature: The south of Europe, especially Northern Italy, remains a pole of great creative ferment in the areas of lifestyle, design, fashion and architecture. This cultural vitality is crucial to the studio in Como in order to predict and anticipate future trends. Moreover, it is not to neglect that the Italian market is one of the most important commercial outlets for Mercedes-Benz after the German market.

The choice of suitable premises where to set up the studio fell on Villa Salazar, facing the Como Lake. A villa commissioned by Della Porta family in the late eighteenth century, completely frescoed, expressing an elegant and refined environment, great inspiration for a creative studio. The Italian centre is responsible for the testing of new interior solutions, while other sites concern is primarily on automobile exterior bodies. A separate room is devoted to brainstorming. A library provides access to records of books and magazines, including fashion and lifestyle. Another area is intended for the choice of leathers. In the villa, there are also tools for reverse engineering and virtual modelling. The design centre of Como comprises a team of 15 people, headed since 2006 by Michele Jauch-Paganetti. His job is to seek solutions of new forms, colours, materials and finishes for future cars of all brands of Daimler AG. The working group is composed by designers, architects, technicians, modellers and artists with heterogeneous training and functions. They make styling and design proposals in parallel and in competition with the other international advanced design studios.

Gorden Wagener, head of design of Mercedes-Benz from 2008, has said: *The strong expansion of our studio in California demonstrates the growing importance of design in Mercedes-Benz. With our offices in California, Japan and Italy, we have established a creative basis for important markets. They function as seismographs for the influences emerging from art, culture and architecture*. The centres of Advanced Design, located in Carlsbad, Yokohama-Tokyo and Como produce independent projects. The proposals are then received by the Design Centre in Sindelfingen, Germany, for final selection and processing.

Case 2: FIAT Advanced Design

Innovation is constant in the automobile industry and because of the time required for the development of an automobile, when planning new products is necessary to implement long-term strategies. The evolution of needs and habits of users, the technologies development and their applications, produce design trends that are constantly changing over time. These factors evolve independently; they must be placed in relation to each other to derive useful and consistent indications. The Advanced Design Fiat is born with this idea: *that of an integrated process, which should be provided as initial nourishment before the stages of design traditionally regarded as project moments (starting 120 months before the completion of the product). It carries structured information about what is happening in the world and how it is possible to translate these useful suggestions locally, for the definition of products* (Advanced Design magazine, Milan 2004).

Fiat identifies Advanced Design as *a center for collection and processing of information useful to the project that will be a doubly functional connective tissue: outside the company, acting as a receiving antenna for the results of research, technological trends, guidelines for finding solutions, to be seized as soon they loom, or inward the company, encouraging exchanges, contacts, circulation of ideas* (Advanced Design magazine, Milan 2004).

Among the tasks identified for the operational structure of Transportation Advanced Design, we can quote: defining plausible scenarios of mobility and its future development, listening and interpreting the signals that come from users and engineers, combining the requirements of engineering with these of product development. Basic objectives are: to give a contribute to the development of new concepts, to harmonize requirements and time with design and production, to act as an interface and a reservoir of ideas, in order to improve final product quality.

The Advanced Design Fiat takes place in parallel on three levels: that of product concept (Concept Lab), the idea of mobility and automobile intended as industrial product (Automotive Lab) and that of materials (Material Lab). There are three separate divisions, although related and complementing their activities, which give to these specific aspects of research their contribution. The research of the three divisions is usually oriented on a common goal and examines what could be the impact of new scenarios on the design of actual production car. For example, the working groups of Advanced Design have already faced project issues in relation to environmental protection and to the new safety standards. To better understand how this could affect the design choices, the complete vehicle life cycle has been studied, until its disassembly for recycling of materials, all the way to the control of waste arising from dismissing.

As a part of the Advanced Design, Fiat in 1998 established the Advanced Design University Stage (ADUS), a programme of activities involving young designers interested in automotive design from the early years of their specialization. ADUS, activating an internship programme for young designers coming from around the world for a significant period of time, made it possible to get a double benefit. For the designers, the in-depth evaluation of practical problems related to design of complex products with a professional experience; for the industry, the possible contribution of young people with the necessary freedom and freshness of mind to solve problems. The ADUS has been openness to new creative resources, oriented to the needs of industry. The goal, for Fiat was that to *prepare human and creative assets for the company's future, providing a permanent strategic resource set to improve the overall product quality through the use and improvement of the talents of young designers.* ADUS has also helped to increase the visibility of Fiat in the design world. In the first 6 years of activity, ADUS has established a network of international relations in the field of education: 258 young designers from 25 countries of all continents and coming from 56 different schools attended its internship.

The relationships established by Advanced Design and, in particular, by Concept Lab, have proved a unique opportunity to test the field difficulties and opportunities of cultural diversity. The research work of Concept Lab has benefited of the contribution of professionals like Isao Hosoe, Alessandro Mendini, Jean Nouvel, Michele De Lucchi, Toshiyuki Kita, Mario Bellini, who have enriched research of innovative directions of undoubted value. These highly qualified designers have provided a positive contribution to the issues identified by Concept Lab and ADUS, thanks also to the experience gained in different contexts from car design.

References

Advanced Design Magazine: Dossier Fiat, Milano (2004)
Arielli, E.: Pensiero e progettazione. Paravia Mondadori, Milano (2003)
Chiapponi, M.: Cultura sociale del prodotto. Feltrinelli, Milano (1999)
Crea, N.: ADV. DGN. Advanced design DiTAC, Quaderno 13, Sala editori, Pescara (2002)
Del Bono, E.: Esseri creativi, Milano, Ed. Il Sole 24 ore (1996)
Del Bono, E.: Creatività e pensiero laterale. Rizzoli, Milano (1998)
Fletcher, A.: The art of looking sideways. Phaidon, Londra (2001)
Vitta, M.: Il disegno delle cose: storia degli oggetti e teoria del design. Liguori, Napoli (1996)

Chapter 8
AdvanceDesign for Product

Raffaella Mangiarotti

> *In essence, what works of design and architecture talk to us about is the kind of life that would most appropriately unfold within and around them. They tell us of certain moods that they seek to encourage and sustain in their inhabitants. While keeping us warm and helping us in mechanical ways, they simultaneously hold out an invitation for us to be specific sorts of people. They speak of visions of happiness.*
>
> Alain de Botton

8.1 Could Advanced Design Be Considered Research?

The term 'advanced' applied to design does not have a meaning generally recognized, so it is used within different contexts to identify high technology design, to identify innovation and to imagine product concepts even away from feasibility.

In this text, with the term 'advanced design' we intend to conceive, problematize, contextualize and finally display research solutions that can predict and change the traditionally conservative path and the incremental evolution of objects.[1]

Advanced design, according to this interpretation, has to do with research and innovation.

According to Archer (1995), 'Research is a systematic inquiry whose goal is to communicate knowledge'.

It is therefore an activity:

- systematic, because it is pursued according to some plan;
- an enquiry because it is seeks to find answers to questions;

[1] With the traditional evolutionary path of the objects we refer to sequence variants, small increments functional and aesthetic renovations of different capacity that characterizes the product development.

R. Mangiarotti (✉)
Dipartimento di Design, Politecnico di Milano, Via Durando 38A, 20158 Milano, Italy
e-mail: raffaella.mangiarotti@polimi.it

M. Celi (ed.), *Advanced Design Cultures*, DOI 10.1007/978-3-319-08602-6_8

- goal-directed because the objects of the enquiry are posed by the task description;
- knowledge-directed because the findings of the enquiry must go beyond providing mere information;
- communicable because the findings must be intelligible to, and located within some framework of understanding for, an appropriate audience.

Archer also identifies three different forms of research:

- research in the science tradition;
- research activities in humanities tradition;
- research through practitioner action.

Science research aims to produce explanations, while research in humanities and arts, typically subjective in nature, has the task of 'evaluating'.

The search for design moves between these two fields and, in relation to the various contributions, can be distinguished in:

- Research on design, which can be of various different kinds, if history belongs to humanities, where the users belong to the sphere of social sciences and so on;
- Research design, which collects a series of studies aimed at improving the practice of design;
- Research through design, conducted through the designers' work, industrial research centres, that is, in effect, a new knowledge.

This form of research is expressed through a series of projects that constitute a 'tacit knowledge', a form of knowledge that is not separated from the perception, judgement and skill which the knowledge informs. This knowledge is transmitted when it is published and 'can result in significant changes in perception and values of persons' (Archer 1995). With regard to this issue, Archer argues that research can be considered only a survey that has the knowledge as objective, systematically conducted in a transparent manner, with results validated in an appropriate manner.

8.2 Which Reference Model?

It is not clear how this could be a general model for design research. Richard Buchanan argues that no one seems to be really sure what design research means. According to the author, design—rather than following the traditional models of academic disciplines—should develop its own model that considers the theory, practice and even the world's production, which is its main objective.[2]

[2] "No one seems to be sure what design research means. Should design research follow the model of traditional academic disciplines, or should it seek a new model, based on the intimate connection among theory, practice, and production that is the hallmark of design?" (Buchanan 1996).

During an international conference on design research, held in Helsinki in 1998, a further distinction is made between design research and design studies, which helps to remove confusion in this field. In this regard, Victor Margolin stands by the research design studies, which have, as their objective, the practical interpretation of design from a perspective of critical theory and humanistic investigation, very different from research-oriented project.[3]

It is therefore evident that the parameters of a design study are not the same of a project-oriented research. Project-oriented research, targeting a product that is an expression of material culture, should not necessarily be accompanied by methodological principles, because, while research has the word as a means of expression and a theory as the objective, design has the material as an expressive tool and the product as an objective.

8.3 Objectives of the Practice-Oriented Research

Design is, therefore, a syncretic and material expression of the culture of our time and thus the purpose is already achieved in the physical production of the object, which must pass its own sense through a 'material narrative'. And therein lies one of the intrinsic qualities of design. An object is considered a qualitative one when interprets the becoming of time[4] through an expression of meaning that must be materialized in the product through a coincidence of individual evolution expressions: technological and material, anthropological, cultural, conscience one and social responsibility.[5]

The product is then, in the first instance, a synthetic and material expression of the social culture.[6]

This close relationship between design and material culture, as cause and effect in nature, becomes more complex when goods are ripe and abundant, technological

[3] "I prefer the term design studies to characterize my own conception of basic inquiry in design as distinct from design methods or project-oriented research [...]. Design studies is an interpretive practice, rooted firmly in the techniques of the humanities and the social sciences, rather than in the natural sciences" (Margolin 1998).

[4] This normally occurs when the design explores the borderlands, the latest technical developments, materials, avoiding the reproduction of what is known and consolidated.

[5] More than half a century ago, Moholy-Nagy (1947) in Vision in Motion, emphasizes the importance of professional responsibility of the designer: 'Thus quality of design is dependent not alone on function, science, and technological processes, but also upon social consciousness'.

[6] The historiography of the French Ecole des Annales has made it known already in the sixties, that if the material culture is the set of instruments and objects produced by a community for a variety of uses, from those related to subsistence to those with ornamental purposes, artistic or ritual, the product is a syncretic way by which the culture expresses and transmits his knowledge. Generally, from the techniques, aesthetics, features, materials of an object can be traced back to the civilization that produced it: because, as Franz Boas understood, it is not possible to regard an object detached from the context that produced it, without immediately cancel its meaning.

offer becomes endless, and territories become global. It would appear that, whereas before the object was oriented to meet the needs of ornamental and/or functional, and express, almost as induced effect, the culture, today, in a world where the values of use and functionality are given, it has become a priority need for cultural, symbolic and communicative expression. It would seem then that design does not only produce innovation in material culture, but that media culture seem to produce design. As already understood in the end of the 1960s Debord in his famous essay, The Société du spectacle (1967), those goods would be organized in 'communicative commodity form' making of their own representation and self-celebration the reason of their existence and of coming consumed.

8.4 *AdvanceDesign*: Technical versus Symbolic

As functional and technological innovation, the meaning is therefore a major goal of design. And this is increasingly evident in the projects of our last generation. While it is easier to assess advanced what is enclosed within a technological or functional contest, since it belongs to the 'technical' and is therefore rational, it is much more difficult to assess what is advance in terms of semantic and meanings, because much more questionable.

In the first case, it is clear that the application of a new technology, a new material, a new function, of a certain importance and with results of common sense, defines an advanced product. Or more clearly, let us imagine to apply a technology that is not available at present, but could become so in the future.

Is it more difficult to assess *advance design* in terms of meanings? For example, the juicer by Philippe Starck for Alessi has made it very clear that the symbolic value of an object can be compared to the predominant functional content. Despite the fact that the juicer does not work perfectly, it has had a huge commercial success and has influenced design since.

As wrote Verganti (2009) in Design-Driven Innovation, innovation is 'design-driven, not by the market, but it creates new markets, does not push new technologies, but gives rise to new meanings'.

Ettore Sottsass was very clear about the market research and marketing departments of labour, saying that marketing was the tomb of innovation. In fact, Branzi and Sottsass (2009) says: "[…] Ettore has taught us that modernity was not in the methodologies, but in the freedom of creative thought, against the logic of marketing testified that it was industry that was to change society, but the company had to change the logic of industry', concepts that anticipate, infact, what we have achieved major design-driven companies like Apple, Artemide, Alessi providing a new vision that does not respond to what people want today, as it is already old, but what they might want tomorrow. Companies use design to capture the new, a new meaning to play, listen to music, to see the light, where there is technology to drive, but the relationship with society and culture. The technology is to support the creation of the new, along with other important stakeholders that are part of a network able to reverse the direction and vision of things.

Here *advance design* does not anticipate scenarios, does not involve consumer research, leverages-proven methodologies, not research, but rather appropriates the 'antennas' of the designer to anticipate new and create emotions. There is not a temporality: in the mind of the designer of the future must already be present and, as said by Ponti (1957) in the book 'Amate l'architettura', in a designer's head exists 'only the present, in the representation that we have of the past and the future foreshadowing'.

Like many other excellent designers, Ponti sees that the product quality, excellence should already foreshadow the future. Quoting Marco Zanuso, 'the project means anticipation, and as such, involves first a reference to the future' (Melograni 2004).

It is important to stress this fact because it is the distinctiveness of what can be considered to be *advance design* in the short term: the ability to turn the vision of things with technology dates. We must recognize that the advance design in the short term is no less important or less influential than the long term. Indeed, we should admit that almost all the foreshadowing that have interpreted the future with a certain distance—and this not only in design but also in the filmography and other arts—have become a source of inspiration for the design trends, but almost never made.

The advance design, therefore, has a deep relationship with the qualitative aspect of the product, a relationship with the expression, with the structural innovation that leads to the new form.

When we speak of innovation, we should emphasize the fact that the term for the discipline of design, has a different meaning from that which can be found in economics, for which identifies the implementation of a creative idea made so as to generate profit for the company, and is distinguished from pure invention, a process that identifies the other hand, an object or a technique that has elements of novelty (originality), but that is not necessarily provided or applied.

In our case, there are objects that have become 'icons' of design without entering the market or having been particularly successful. This is related to the fact that, to become an icon, it is sufficient that the image of an object to be disclosed in a consistent way through the media. And this is especially evident in several cases of products not only sold a little, but also actually remained at the prototype stage and never become a part of the market and 'ever produced'.

For example, the radiator designed by Joris Laarman, after being published in magazines and have won design awards, it remained a prototype as it was not really easy to produce on an industrial scale. Given the success of image, only recently Jaga, a company in the specific field has decided to put it into production, albeit with almost artisan methods.

This case has shown clearly what the consumer and the press seem attracted by the potential of a decorative object, and paved the way for a series of imitators who have used the same concept to create similar products, a phenomenon that has profoundly changed the idea of a radiator.

So it seems that this process of diffusion of the product image, often regardless of whether it is still a prototype, because of its narrative power and evocative stimuli can fertilize new ground that gives rise to the birth, almost as a fashion effect, a series of side products, which seem to result in a natural manner by the strong product of reference.

Without media coverage, this process would not be possible. Until a few decades ago, in fact, the marketing, distribution and commercial success were the only main factors to influence the evolution of products.

8.5 Linguistic and Formal Innovation

The theme formal and linguistic innovation triggers always a hot debate. Many argue that the formal and linguistic experiments are not a genuine kind of innovation. This is related to the fact that the design discipline has traditionally based its founding principles to an intelligent relation between function and shape and design, working a lot on the concepts of simplicity and minimal form.

In fact, the production environment has changed so much in recent decades. New technologies informing designing and developing process can be used both to prototype and to finish products. This has changed the way people think to design and to project.

A strong example of change is in the rapid prototyping technologies, in the past not so used because of high costs. Today, lower costs have given them the potential for future employment as real production techniques. This has given birth to a series of prototypes that have created new market niches.

Among the several case studies, the work of industrial that due to diffusion through the important and sophisticated media channels and some design awards worldwide, has helped spur new languages in the world of the accessories.

Nothing different from what had already happened with Memphis, but then the gallery system and the spread of the Internet sales have created channels alternative to traditional industry production.

8.6 The Term Advanced Has a Relationship with Innovation and with Time

Design is a temporal practice: to design is to project objects or services available to the market at a predetermined time and make sense precisely in relation to that moment. The understanding of the right time for the introduction of a certain innovation into the market often defines the product success itself would introduce a distinction between the designer and the company.

Normally, while in the mind of a designer to design, as we have said, has the intrinsic characteristics of innovation and a presentiment of the future, the world of enterprises, especially large enterprises, it needs a very long time to introduce innovation, requires schedule from small incremental improvements to great innovation.

Normally when it comes to advanced design, it refers to a medium-long period of time, because companies need to understand the directions of innovation compared to sociological and cultural change to orient their products and their research.

This time lag may be from few to 10 years in the case of individual products, but also extends to 20–30 years, when it comes to more complicated topics such as those pertaining to traffic management solutions, or energy problems, etc.

Generally, this kind of work is done in a team and relies on the construction of scenarios. The scenarios provide a basis and a framework, a tool of cooperation within the design team, but also a door opening outside professional boundaries, and to built interdisciplinary teams (sociologists, cultural anthropologists, cognitive psychologists, technicians, ergonomists, etc., often in teams are also actively involved stakeholders and potential customers).

The scenarios are used to build a shared vision about the project and give rise to products, expression of the material culture of the scenario.

Advanced scenarios are typical of large enterprises and are used for several purposes:

- to communicate to the market, the company predisposition to product innovation and attention to the future;
- to share within the enterprise, the scenarios related the evolution of products or processes;
- to iconify the innovation forms. This is one of the strongest ways of innovation propulsion, causes the disclosure of a new idea is a way to share the typological evolution of the product even outside the company and colonize territories yet unexplored;
- to be more visible to the market in comparison to its competitors and free media advertising while reducing costs;
- to increase the design centre skills through reflection on non-routine design issues, participating and exchanging ideas, methods and tools with multi-disciplinary consultants and external designers.

In this context, there are several case studies: Vision of the Future scenarios from Philips in consumer electronics field to those of Whirlpool appliances.

Case 3: Philips Vision of Future

Vision of Future was a research led by Stefano Marzano's about desires and needs in the field of information technology, entertainment and communication from the company in the near future. The survey was projected 15–20 years and was born from a process of imagination and processing of data coming from the socio-cultural and technological opportunities. The new interpretation has led to four scenarios: the area around the person, of domesticity, public and mobility.

The process involved the creation of 370 different scenarios by an interdisciplinary team. The scenarios were then subject to a validation process by international opinion leaders who study the world and the idea of the future with a scientific and humanist approach, whose revision has resulted in a selection of just 60. Even abstract scenarios were defined at a higher level of tangibility, to be conducted at a second step of validation by a wider audience. In this phase, prototypes have been made but even if non-existing, not working objects, able to communicate a new material, which were then told in short filmic narratives highly interactive that communicated what they were, what they did and how they did it.

The creation of prototypes and interactive stories has helped the process of understanding and acceptance of the new: the materiality of the proposals was so real that the audience could not tell what it was a real product and what was only a prototype. This method gave strong credibility to future products and defined very interesting feedback from consumers for marketing research and innovation acceptance.

Case 4: Whirlpool Europe Project F

A similar experience to that of the Vision of Future was conducted by the group's Global Consumer Design (GCD), the centre led by Richard Eisermann, responsible for design research, strategy development and its applications to product design in Whirlpool Europe.

The work of the GDC covers three levels:

- A level of design research, whose goal is not necessarily to create products, but to investigate possible ways of interpretation and new paths of innovation. At this level, the purpose is to develop such a high level of innovation that can build a path to guide innovation in the enterprise;
- A strategic level, starting from the previous phase research, identifies potential products to be developed in a time span ranging from 5 to 10 years;
- A tactical level, over a period ranging from 6 months to 2 years, which responds to rapid changes and tastes of the market by offering the most timely product design.

These three levels influence one another and draw energy from one another. At the end of the 1990s, the GDC begins a series of exploratory projects designed to advance it is going to understand and meet the future needs and desires of consumers. The first of these was in 1999 and titled Microwaves, dedicated to the exploration of new ideas on the future of microwave ovens.

A second project, Project F: Fabric Care Futures, began 2 years later and aims to investigate the future of textiles and clothing care through the washing to innovate the classic white box. For this project, the GDC will search for the traditional expectations of markets and a series of interviews with consumers, investigating in particular aspects of use and ergonomics. Alongside this practical research, was commissioned at Future Concept Lab of sociological research—New Domesticity—in order to understand the changes in the size of domesticity in Europe.

The project objectives were: to understand future needs and desires of the consumer; to propose new project themes for the new products development; to cause an ideas and expertise transmigration among the designers and interior design consultants teams, to communicate the research to market.

The study lasted for about a year, from the definition of research, the brief preparation, the concept and prototype definition.

The Project Workshops

The research results were used to provide basic information useful to designers. Three design studies are in fact called to give a reading international Designkoop (Germany), Deepdesign (Italy) and Designraw (the United States), to which is added an internal team of six designers.

Designers are invited to attend a two-day workshop in a beautiful and alienating environment of Varese, where they are informed about consumer trends, washing technologies and about qualitative and quantitative research results.

The process starts by identifying a range of general interest topics: sensory, relationship with the domestic space, ritual and social interaction, attention to fabric and textiles of the future, environmental and ecological issues. Within these themes the designers start brainstorming, formulating potential ideas that can pose problems and solutions within the general themes. Brainstorming is done through the 'method of post-it', in which each designer writes considerations, problems and potential solutions within large sheets hung on the walls, which focus the various issues of interest. This development process is extremely interesting because, besides the production of individual designers, creates interpersonal stimulation.

A phase of discussion and performance monitoring and scenarios visualization, defining issues and possible directions to take share and possibly give.

After a month, working groups have a number of concepts at GDC, including defining what is the most interesting to develop for each designer.

In the following month, it is defined as a further development of the concept which also includes the development of three-dimensional engineering drawings in order to perform a dimensional model.

It then organized a second workshop to 2 months apart, during which it presented a further step forward. The presentation of the products takes place in front of an international management at Whirlpool, in the Hall of Explorers of the evocative Villa Ponti in Varese.

Only from this phase, the Whirlpool internal design team started to develop their concepts.

Ideas Prototyping

The next phase involves the creation, within the next 2 months, three-dimensional solid models by Whirlpool, except for the Deepdesign model which is implemented directly by the designers with the help of a model maker, following implementation difficulties. The five projects offer a rich and varied scenery of proposals for the fabric care, where not only the types of solutions are very different, but equally refer to different temporality. For example, Pulse of the project and the project Deepdesign Bodybox of Designkoop are immediately feasible, and projects of the group Om Biologic and GDC are still subject to a necessary technological evolution.

All projects are still based on a formula basic technique: container + water + soap + agitation = clean.

Each element of the formula is carefully examined: the size, shape, proportion and arrangement of the containers, the necessity (or less) of water and/or detergent and alternatives mechanical agitation (rubbing, dipping, compression, separation of fibers, rotation, vacuum, hydrodynamics, vortices and jets of water).

The group developed the concept Designkoop Bodybox, a highly technical piece of furniture that combines body-care and care of the fabric. The machine assumes a selection of the heads, and a distinction between them for the washing process made on the basis of the reading of the labels. At the same boxes are provided cromoterapic lamps and a shower for relaxing the body.

Pulse of the project Deepdesign assumed an organic form inspired touch to the body, the delicacy of touch that emulates the behaviour of human hands replaces the centripetal force to the centrifugal pump simply takes the place of the old and noisy engine. The pump controls a flow of air that operates the rubber membrane which, in turn, expands and compresses on the tissues, as do the hands by eliminating the engine noise and vibration placed at the sound of water. The machine is similar to a heart that massages the tissues with a new concept of care has a mild form and a soothing sound and thus can be accepted in unconventional spaces of the house as the living area.

Cleanscape is the project Designraw. It is a washing machine that was inspired by the memory of the washerwomen who wore the clothes on the bank of the river, reinterpreting the form of primitive elements of the ritual of washing the mortar and pestle. The idea is to socialize the time of washing: the user brings his basket of laundry in a common place within environments such as bars fitted out with comfortable sofas.

The GDC team has developed two projects. The first, Om, is a machine programmed to mimic the operation of the enzymes. Based on nanotechnology, it removes dirt, removing and replacing the water detergents. The project Biologic, however, uses the concept of phytoremediation with hydroponic plants for the renewal of the water and fuel cells as non-polluting energy source. The fabrics are distributed no longer in a single basket, but in different containers which accommodate hydroponic plants.
The concepts that have attracted most attention from the press were Biologic and Pulse for the evocative and emotional content.
The next steps for the GDC were to re-evaluate and develop the ideas that have originated in Project F, but according to multiple objectives consistent with the strategic plan. Some project ideas are reviewed and revised in future products and produce fertilizer on other projects over the short term. For example, in later years, some solutions are translated directly from the project, such as the creation of a silicon pockets inside the baskets can strip the linens, but also incorporate the idea of washing machine in a system for the furniture. Project F has nevertheless been valuable to ask important questions. The answers to these questions will influence the future strategies of innovation at Whirlpool.

The Communication

After the prototyping phase, in early 2002, the GDC develops the communication phase of the project, consisting of a series of activities: participation in exhibitions, a publication, a CD ROM and a website that explains the project.
Project F is exposed at Hometech, a trade show aimed at a technical field of public management. Subsequently, in April of 2002, is presented during the Salone Internazionale del Mobile in Milan in the space of Pelota. Following exposures to the Paris fair of commercial Step in the Future (in June) and in Brussels in August 2002.
The project Pulse also won the ID Award for Best Concept innovation of the year 2003, tied with another project developed by the company Ideo.

The Results

The presentation of prototypes and projects in numerous publications in journals and books has undoubtedly helped to promote the Whirlpool brand positioning as an innovative brand and extremely design-oriented. In fact, Whirlpool, after the two projects led by Eisermann, which lasted for less than a year, has acquired an authority in the innovative design of the appliance that has few competitors in the world (probably only Electrolux). The quality of the results has meant that publications come out of newspapers and magazines skilled in design, technology and style. Suffice to say that, in economic

terms, the value of editorial space in magazines and books was equal to the total budget that Whirlpool has made available for research.
Project F and Microwaves Whirlpool Europe have triggered within extremely busy that he dismissed the group's GDC, but also other groups involved in the project, from the usual routine deployment and re-open the imagination and discussion on new technologies. It is clear that the exploration of design came through the work and the perception of the future of design consultants and members of the GDC, Whirlpool has made an important contribution to address innovation and strategies, highlighting new opportunities for the future. The creation of prototypes and their physical and virtual disclosure that arose on the icons of the future is able to respond to future desires and emotions. The disclosure not only to the general public but also within the specialized fairs has led to the opening of a debate on the future of fabric care with stakeholders from different levels. The communication of the projects within the Whirlpool not only in Italy, but also in Europe has also contributed to a dialogue between different business functions on a national and international levels.

References

Archer, B.: The nature of research. Co-des. Interdiscip. J. Des. 6–13 (1995). http://www.metu.edu.tr/~baykan/arch586/Archer95.pdf

Branzi, A., Sottsass, E.: O la libertà del pensiero creativo. Interni **580**(4), 72–77 (2009)

Buchanan, R.: Book review. Elements of design. Des. Issues **12**(1), 74–75 (1996)

Debord, G.: La Société du spectacle. Buchet/Chastel, Paris (1967)

Margolin, V.: The multiple tasks of design research. In: Strandman, P. (ed.) No Guru No Method? Discussion on Art and Design Research, pp. 43–47. University of Art and Design, Helsinki (1998)

Melograni, C.: Progettare per chi va in tram. Il mestiere dell'architetto. Mondadori, Milano (2004)

Moholy-Nagy, L.: Vision in Motion. Institute of Design, Chicago (1947)

Ponti, G.: Amate l'architettura. Vitali e Ghianda, Genova (1957)

Verganti, R.: Design-Driven Innovation. Etas Libri, Milano (2009)

Chapter 9
AdvanceDesign in Lighting

Maurizio Rossi

Over time, everything is changing.

Leonardo da Vinci

9.1 Introduction

This paper intends to lay the investigative basis on the elements that contribute to design, in terms of conditions and practices useful in encouraging continuous innovation, in the field of Lighting Design. In this context, the word *advance* is understood not as the adjective *advanced,* but as a verb that describes a movement, a leap forward towards the future, towards new scenarios of innovation in lighting design. A key element of this design vision is its multidisciplinary approach, which should not be understood within the limitations of a pluridisciplinary approach, that simply combines different skills, mainly for the purpose of comparison, but rather employs interdisciplinary methodologies that, through mixing different types of knowledge, try to extract a new epistemology that is not just the sum of the competencies working together.[1]

It is in the quest for *know what* that invention, with its basis in multidisciplinary aspects, applies new ideas and new knowledge or acquires them from other areas of basic or applied research to allow lighting design to advance its horizons. Through research, art, education, practice, communication and other areas, the imagination is stimulated to allow lighting design to be cast into the future.

[1] In this regard, please see Callari Galli and Londei (2005).

M. Rossi (✉)
Dipartimento di Design, Politecnico di Milano, Via Durando 38A, 20158 Milano, Italy
e-mail: maurizio.rossi@polimi.it

M. Celi (ed.), *Advanced Design Cultures*, DOI 10.1007/978-3-319-08602-6_9

9.2 Advance Lighting Design

For several years, lighting design has no longer been a field dominated solely by technical aspects encoded in the rules, standards and laws of lighting. Just as lighting is no longer just a question of photometric evaluation of light by human beings, because light radiation has broader effects on individuals than just its visual function, as evidenced by Rea (2008), Director of the Lighting Research Center at Rensselaer Polytechnic Institute, and Boyce (2003).

With Advance Lighting Design (ALD), the aim is to envision the design horizons of how the relationship will be between individuals, light and spaces, with a view towards environmental sustainability. It is not only a science or an experimental art, but covers both of these two things, and integrates them with the imagination of human beings in a holistic view of ways of thinking, with the aim of improving quality of life, while taking environmental issues into account. This kind of holistic design vision is opposed to the trend towards the super-specialization of some design groups and prefigures a fusion of different (and changing, depending on the context) types of knowledge and expertise, in order to stimulate the imagination with regard to the technical, aesthetic, ethical and economic characteristics of the project. In fact, specialization in one area alone makes it more difficult to approach and view the problem in such a way as to integrate different and complex elements in a comprehensive design solution.

In introducing ALD, we are therefore not putting emphasis on lighting (Bonomo 2000), but rather on light as a fundamental element in relation to the human element, and taking environmental issues into account. In this sense, ALD does not exclusively focus its attention on the design of the lighting system or the lighting product, but on the combination of these two things, in relation to human beings, trying to foresee their future needs and desires. This goal is pursued on the basis of scientific discoveries, the development of design culture and artistic research related to light.

In this attempt to envision scenarios for the future, ALD is moving from the rigid path of consolidated and codified knowledge, towards the sometimes uncertain horizon of experimentation, reasoning both in terms of objects/systems and processes, to respond, as claimed by Paolo Inghilleri (1999), to a wider set of needs of individuals than just those related to visibility, and prefiguring design solutions in which hybridization, with different themes from lighting alone, is already an accepted fact, as evidenced, for example, by the development of the products of the Air, Light, Sound, Other object (A.L.S.O.) line by Artemide in 1996 (Fig. 9.1).

The elements that contribute to developing the creative thinking processes in ALD are based on multidisciplinary design research, on experimentation in teaching (also through design workshops) and on scientific and cultural communication. These are then conducted in the foreshadowing of innovative design action in terms of consultancy, product designs, systems and processes and the development of design and production networks.

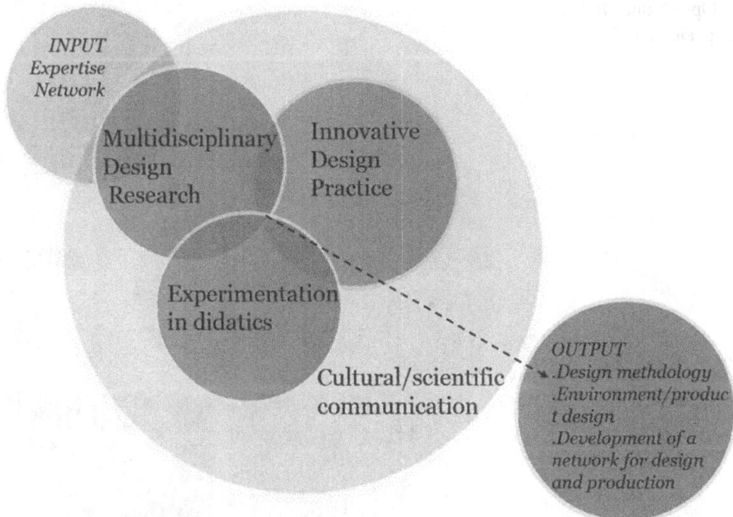

Fig. 9.1 A proposal of the elements/actions involved in the design process Advance Lighting Design (ALD)

9.3 The Key Elements of the Project

When we talk about lighting, the four elements around which design practice rotates are human beings, light sources, the surfaces of the environments to be illuminated and the ecosystem.

Historically, sources and surfaces are the main elements for which lighting has developed, in which the individual is only contemplated as a being endowed with the sense of vision and therefore deemed a 'photometric being'. However, human beings, with their biological organism, their needs, aspirations, desires and emotions, cannot be simply described by photometry. Other fields of science and other disciplines come into play. Next to radiometry and photometry, we should in fact consider the knowledge derived from physiology and psychology and knowledge that is inspired by artistic and emotional experiences. In recent years, a school of thought has developed in opposition to the design of beautiful lighting products as ends in themselves. Rather, it favors the development of a design process that begins with an analysis of the environment and of the activities that humans carry out there, to arrive at a foreshadowing of how the light should be in this environment, and only in the final analysis phase, defines how the product/system that produces this light could be (De Bevilacqua 2008).

Finally, the environmental compatibility and the eco-sustainability of the project are inescapable elements as they now come into play in the standards and legislations of this field—though not always appropriately—witnessing a new culture of respect for the environment, understood to be a limited resource (Fig. 9.2).

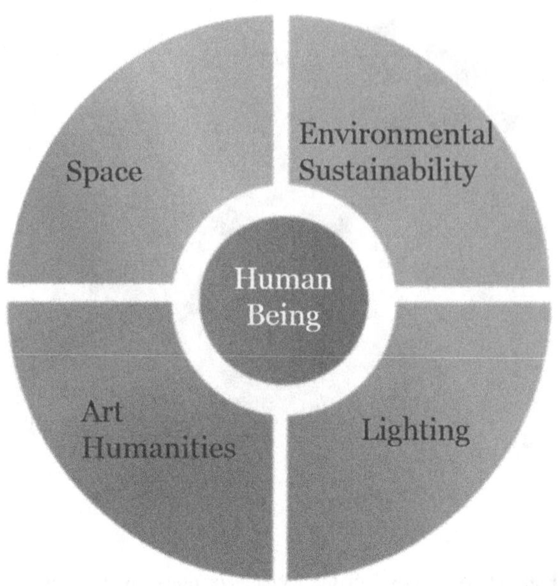

Fig. 9.2 Operational fields of Lighting Design

The development of design sensibility towards issues of environmental sustainability has long assailed almost all of the design discipline. Eco-design is based on the key concept according to which the environmental compatibility and the sustainability of projects are basic requirements throughout the entire design process (Manzini and Vezzoli 2007; Yeang 2004). Referring to any type of project, Eco-design, as applied to lighting, is based on a set of principles and follows certain rules and various approaches.

The process that follows the design of an object "from cradle to grave" and later evolved into the concept "from cradle to cradle" (Mcdonough and Braungart 2002), takes into account the entire life cycle of a product, until its rebirth/re-use, considering its potential impact on the environment. The Life Cycle Assessment (LCA) is, in fact, a useful method for the objective assessment and quantification of energy and environmental loads and of the potential impacts associated with a product, process or activity, from the acquisition of raw materials to their end of life. Furthermore, it also considers, while assessing the environment impact, the stages following disposal, namely the possible ways of recycling and reusing all or some components of the product.

Other key elements are: reduction in resource consumption, reduction of energy consumption, reduction in the use of pollutants, use of renewable, reusable or recycled materials, reduction of post-consumer waste and of garbage and better waste management.

9.4 The Basis of Advance Lighting Design

The relationship between research and teaching, development of design and development of the designer, is characterized by the education of the latter and the scientific, cultural and experiential background that he/she brings to the design practice. Just as Basic Design (BD) stands as a central foundational discipline of design, of which it tries to balance the aesthetic, technical and scientific components, Basic Lighting Design (BLD) (Rossi 2008) stands as a basis for ALD, where it experientially tries to bring out the physical aspects of phenomenology, the visual-optical and perceptual-cognitive ones, which are the foundations of the design relationship between individuals, light and space.

A first attempt to frame the multidisciplinarity aspects that contribute to ALD thus shifts the steps from the four basic elements of this project framework (individual, light, space and ecosystem), to explore others, whose instances arise through the relationships between the main ones (Fig. 9.3).

With reference to individuals, physiological, psychological and sociological aspects come into play when they are considered in connection with the space in which they find themselves and with other individuals. The physiological aspects relate to the theme of well-being in relation to the circadian rhythm that regulates the biological clock of our body under the influence of light. The psychological aspects adhere to research about the emotional states and moods, depending on the quality of light (Rossi et al. 2006, 2009). In between the physiological and psychological aspects, and in relation with the light and space, there is the issue of visual perception. This, in fact, is not a mere collection of information or light signals that we receive from the environment, but rather it is the cognitive ability to process and perceive the visual information received (Gregory 1997; Frova 2000).

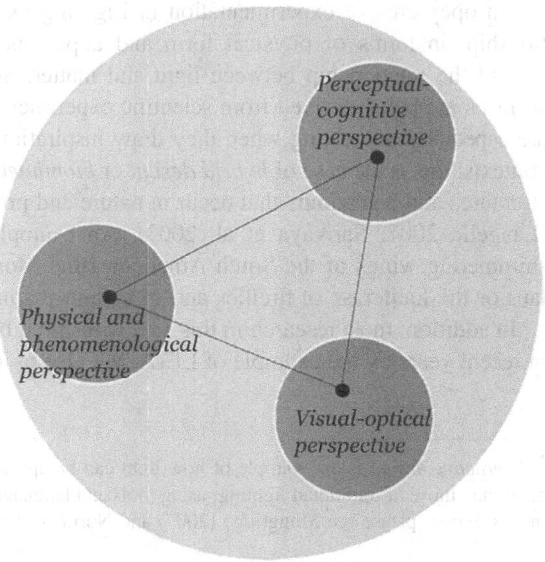

Fig. 9.3 Fundamentals of basic light design (BLD)

This is an open research topic in various disciplines, but from an experiential point of view, a knowledge of BLD is essential for designers. Among the sociological aspects, we must surely include spatial anthropology with its ability to investigate the individual with respect to the space, to the contexts of use and to the ways of perceiving and experiencing the space (Virilio 1984; Augé 1992). Of particular interest is proxemics, which investigates the relationship between the perception of personal and social space in relationships with other individuals and within an environmental, cultural and emotional context (Hall 1990).

Around the individual and the space, there is also the artistic experience in which light is involved. It is in this context that issues related to culture, inventiveness and emotion find fertile ground. A first aspect concerns the evaluation of the role of light in the work of art for which a collection of writings by a popular art critic, are of certain interest: Hans Sedlmayr. He deals with light also in scientific terms, but without the intention of implementing a systematic discussion of the phenomenology of light. Rather, he pays attention to the cultural question of light, trying to describe it by developing an appropriate taxonomy (Sedlmayr 1979). Another very interesting aspect is that in which light itself becomes art through light installation, which in most cases is temporary and for that reason, easily forgotten (Weibel and Jansen 2006; Gellini and Murano 2009).

If we consider light, we should definitely contemplate its physical and technological aspects. Also in this context we recall that part of BLD that addresses the physical phenomena of light. Research and skills in this field relate to light sources, lighting equipment, materials and methods of designing products and systems, which arise in connection with the issue of sustainability.

Experiments with materials open up unexplored perspectives into the generation of innovative projects. Very interesting examples of this hybridization of skills, disciplines and sectors can also be seen in projects that are generated in areas outside Lighting Design.[2]

An open area of experimentation in Lighting Design is that of the direct relationship, in terms of physical form and appearance, which is being developed around the relationship between light and matter. Some innovative materials and surfaces are being derived from scientific experiments in nanotechnology, but they are especially interesting when they draw inspiration from distant and unexplored contexts: this is the case of *hybrid design* or *biomimetic design*, which is inspired by structures and behaviours that occur in nature and proposes them as design projects (Langella 2007; Sarikaya et al. 2003). An example is the pigmentation of the shimmering wings of the South American Blue Morpho butterfly, Morpho menelaus or the luciferase of fireflies and of certain marine animals.

In addition, most research in this area is inspired by new light sources, as shown in recent years by the example of LEDs, but also by legislative and systemic issues

[2] *Interactive design* is an example of how light can be applied according to ways and purposes other than those of traditional lighting, using tools and languages that are entirely original at times. In this regard, please see Moggridge (2007) and Norman (2002).

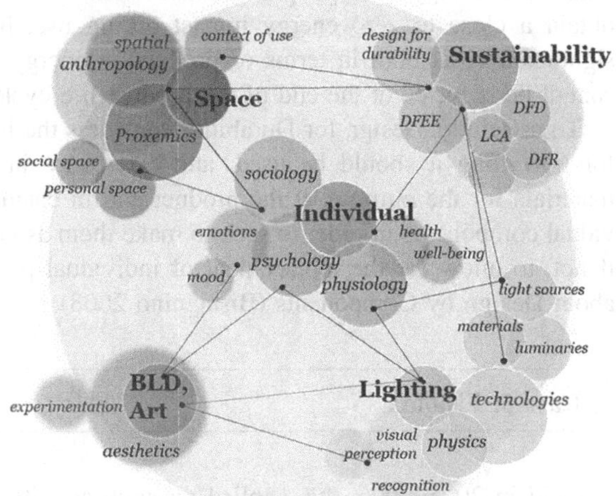

Fig. 9.4 An interpretation of relationships and multidisciplinary elements that foster ALD

related to the banning of certain products or by compliance with sometimes controversial laws, such as the one on light pollution, where the term "pollution" is judged inappropriate by many people (Bonomo 2000).

In lighting, Eco-design applied to the object and to the system is affected by various areas of design: Design for disassembling (DFD) considers how to disassemble components to replace, recycle and reuse them; Design for Recycling (DFR) considers whether the individual components and materials are recyclable, while Design for Remanufacturing (DFM) investigates the possibility of replacing a part or reproducing a component. In this particular context, it is important to consider the life span of an object and the combination of different life spans of individual system components (Fig. 9.4).

For the aspects of design concerning process, Design for Energy Efficiency (DFEE) tackles the problem of sustainability in terms of energy consumption by taking into account the entire production process of the components. In particular, a product, during its use, is not necessarily impacting in terms of energy, but it is often true that during the production process, the amount of energy consumed to produce the various parts is very high.

In this case, the production process must be maximized and optimized for better energy efficiency. For example, in the realm of light sources, those that appear to be efficient in the production process, such as traditional incandescent bulbs, are however very impactful in the life cycle of use. On the contrary, sources that seem ecologically sustainable for their low power consumption, such as compact fluorescent bulbs, are characterized by particular problems during production because of the materials used and by further problems with regard to the end of life phase. It is therefore desirable that, in a truly eco-friendly design of product-process, one develops a system for collecting and recycling these sources (as already occurs for linear fluorescent bulbs). The same situation applies to light sources that use renewable energy through photovoltaics that, on one hand make it possible to obtain a close to zero energy impact during use, but on the other hand, have significant costs both in terms of production energy and use of materials, which cannot be recycled at the end of the product life cycle.

To conclude, Design for Durability considers the life span of an object, namely for how long it should be used, and takes care in choosing the most suitable materials for the purpose of the product and in calculating the life spans of individual components in order to seek to make them as homogeneous as possible and, if not, to allow for the replacement of individual parts. In this case, we can talk about Design by Components (Bistagnino 2008).

Case 5: Fisiolux

Held in 2005–2006, this applied research activity was aimed at developing innovative lighting systems in the residential field. The research activity temporally preceded the activities of design development, in order to provide new foundations for product requirements, which included as a priority objective, the conscious use of bio-physiological effects of artificial light on the human body.

Developed at the Light Laboratory (lighting, photometry, colorimetry and perceptology) at the Politecnico di Milano for a leading Italian company in the lighting sector (Artemide S.p.A.), the research has provided design specifications for the creation of a line of products called My White Light. It was also placed in a broader framework of subjects, which involved the cooperation of the San Raffaele Hospital in Milan and psychologists at the University of Milan.

In the field of lighting design, the concept of well-being has been discussed for many years, referring to the quality of vision: an example includes research into direct and indirect glare control, into contrast rendition and into the contrast between the area of the primary and background visual function. Even manufacturers have addressed the issue of quality of vision by improving the colour rendition of fluorescent and discharge sources.

Recent research, conducted in the medical field regarding the effects of light on man, shows, in addition to aspects related to the production and operation of lighting, new standards of quality related to human physiology. Light has a direct impact on the activity of the cerebral cortex, on the body temperature and heart rate. It has been scientifically proven that it can have therapeutic effects by positively influencing the treatment of Seasonal Affective Disorder, sleep disorders, quality of the phases of sleep and wake in Alzheimer's patients and weight gain in premature babies. It is also known to regulate the production of the melatonin hormone, depending on the spectrum and timing of eye exposure. Extensive research has finally highlighted that the human circadian system is regulated primarily by the presence of a photo-reception

mechanism found in the retina, due to a photopigment called melanopsin by some researchers.
Based on these and other studies, Lighting Design can deal with new aspects that affect human well-being, with regard to the design of fixtures and systems. Among these we can mention a control of the amount of lighting, control of the real spectral distribution (colour) of the artificial lighting and the timing and duration of the lighting levels and quality through dynamic, automatic or manual control.
However, these elements must be accompanied by a careful analysis of the type of activities carried out within the areas to be illuminated by artificial lighting. This should be able to adapt to the development of new scenarios of work and space organization, but also to the new structure of personal and collective needs, aiming to strike a balance between psychophysical well-being and efficiency of work activities.

Case 6: Food Lighting

In 2003, while addressing the problem of the best lighting of food in stores, the company Castaldi illuminazione SpA contacted the Light Laboratory of the Politecnico di Milano in order to obtain useful insights for the development of new lighting products that could integrate new technological solutions recently introduced on the market. We quickly realized that we were in an industry dominated by solutions that are not supported by adequate scientific or experimental studies, but instead entrusted to subjective, varied and sometimes quite fanciful judgments. We therefore decided to carefully analyze the relationship between light, food and individuals, researching both the technological issues inherent to new high-efficiency light sources with various colour temperatures, and the psychological relationship between the appearance of food and the choices of individuals, in order to create new services in hues of white light useful in enhancing the different types of fresh foods, which are usually on sale.
The merit of this research, apart from the fundamental theoretical seriousness and the completeness of the metrological measurements developed, is to have involved a large number of individuals in numerous experimental tests. The perceptual tests were performed on a sample of 124 people equally (50 %) distributed in the two sexes and divided into various age groups. The subject, faced with the same foods, illuminated by different spectra, indicated his/her purchasing preference.
The perception of colour and the greater or lesser appreciation of one type of light rather than another, are in fact judgements made by each of us with our own eyes and our own brains. It is therefore very much our duty to investigate them. The real infallible judgement is that of the observer, whose statistic preference has certainly more value than that of many ‘experts’. With this

research, we obtained a double certainty. On one hand, a theoretical certainty and, on the other, an experimental one, with a clear indication of the most appropriate way to illuminate different foods in terms of the quality and quantity of light.

It is known that the predisposition for purchase and consumption of products depends, to a considerable extent, on the appearance of what we are observing; this consideration is even more true for fresh food. The properties of surface appearance (Hunter and Harold 1987) of foods such as colour and brightness, can be emphasized and influenced by illumination levels and especially by the colour-spectrum properties of the light source. The appearance of the food, as it is perceived by the visual system, influences complex psychological phenomena in the cerebral cortex (Zeki 1993), which influence our evaluation of the quality of the product: if it is more or less fresh, ripe or unripe, appetizing and so on. Based on these observations in this research, extensive quantitative metrological and qualitative perceptual surveys were conducted. The foods were chosen from among those which, in the distribution of large shopping centres, are exposed to artificial light either directly or through transparent packaging (or a box) and whose direct observation involves the buyer in a visual and subjective evaluation of the product. The types of foods chosen were five red meats and bakery products, oily fish, fruit and vegetables, green, yellow–green and yellow in colour, and finally orange, red and other assorted colours of fruit and vegetables.

From the results obtained, a line of products were created that, by integrating high-efficiency discharge light sources, can obtain different colour temperatures of white light through the use of dichroic filters (Castaldi et al. 2004). In subsequent years, other manufacturers in the industry introduced, for food outlets, new lighting systems that could control the colour temperature of the light in order to appropriately control the tones of white to be used for various foods.

References

Augé, M.: Non-lieux. Introduction à une anthropologie de la surmodemité. Paris, LeSeuil, (1992)

Bistagnino, L.: The Outside Shell Seen from the Inside, Design by Components Within an Integrated System. CEA, Milano (2008)

Bonomo, M.: Una legge da rifare, In: “Luce”, no. 5, September 2000

Boyce, P.: Human Factors in Lighting. CRC Press, Boca Raton (2003)

Callari Galli, M., Londei, D.: Multidisciplinarietà oggi. In: Callari Galli, M., Londei, D., Soncini Fratta, A. (eds.) Il meticciato culturale: luogo di creazione di nuove identità o di conflitto?. Clueb, Bologna (2005)

Castaldi, G., Fallica, C., Rossi, M.: Illuminazione degli alimenti - Una ricerca applicata. Luce Design **6**, 84–89 (2004)

De Bevilacqua, C.: La luce negli spazi umani. In: Rossi, M. (ed.) Design della luce: fondamenti ed esperienze nel progetto della luce per gli esseri umani. Maggioli, Santarcangelo di Romagna (2008)

Frova, A.: Luce, Colore, Visione. BUR, Milano (2000)
Gellini, G., Murano, F.: Light Art in Italy. Maggioli, Santarcangelo di Romagna (2009)
Gregory, R.L.: Eye and Brain, 5th edn. Princeton University Press, Princeton (1997)
Hall, E.T.: The Hidden Dimension. Anchor Press, Norwell (1990)
Hunter, R.S., Harold, R.W.: The Measurement of Appearance, 2nd edn. Wiley, New York (1987)
Inghilleri, P.: Light Fields. Artemide, Milano (1999)
Langella, C.: Hybrid Design. Progettare tra tecnologia e natura. Franco Angeli, Milano (2007)
Manzini, E., Vezzoli, C.: Design per la sostenibilità ambientale. Zanichelli, Bologna (2007)
Mcdonough, W., Braungart, M.: Cradle to Cradle: Remaking the Way We Make Things. North Point Press, San Francisco (2002)
Moggridge, B.: Designing Interactions. MIT Press, Cambridge (2007)
Norman, D.A.: The Design of Everyday Things. Basic Books, New York (2002). Reprint edition
Rea, M.: More than vision. Editoriale Domus, Milano (2008)
Rossi, M.: Design della luce: fondamenti ed esperienze nel progetto della luce per gli esseri umani. Maggioli, Santarcangelo di Romagna (2008)
Rossi, M., et al.: On light and color effects for interior lighting Design: part I theory. In: Rizzi, A. (ed.) Colore e colorimetria: contributi multidisciplinary vol 2. II Conferenza Nazionale del Gruppo del Colore SIOF, Milano, September 2006. Collana quaderni di ottica e fotonica, vol 15, Centro Editoriale Toscano, Firenze, p. 145 (2006)
Rossi, M., et al.: From physiology to a new sustainable lighting design: the "My white light" case study. In Bertram, P., Botta, M. (eds.) Multple Ways to Design Research—Research Cases that Reshape the Design Discipline, 5th Swiss Design Network Symposium, Lugano, November 2009. Et al./Edizioni, Milano, p. 272 (2009)
Sarikaya, M., Tamerler, C., Jen, A.K., Schulten, K., Baneyx, F.: Molecular biomimetics: nanotechnology through biology. Nat. Mater. **2**(9), 577–585 (2003)
Sedlmayr, H.: Das Licht in seinen Kùnstlerischen Manifestationen, Màander Kunstverlag, Mittenwald. It. transl.: Albarella, P. (ed.) (1994) La luce nelle sue manifestazioni artistiche, Aesthetica edizioni, Palermo (1979)
Virilio, P.: L'espace critique: essai sur l'urbanisme et les nouvelles technologies. Éd Christian Bourgois, Paris (1984)
Weibel, P., Jansen, G. (eds.): Light Art from Artificial Light, Light as a Medium in 20 and 21 Century Art. Hatje Cantz Verlag, Hamburg (2006)
Yeang, K.: A Manual for Ecological Design. Wiley, New York (2004)
Zeki, S.: A Vision of the Brain. Wiley-Blackwell, Hoboken (1993)

Chapter 10
AdvanceDesign in the Reconfiguration of Relationships Between Companies

Stefania Palmieri

> *Reviewing what you have learned and learning anew, you are fit to be a teacher.*
>
> Confucius

The turbulence of competitive, technological, social and market scenarios is posing companies unprecedented challenge in innovation of products and services. Designs are increasingly characterized by a considerable amount of new content, in spheres in which experience and know-how developed in the past are hard for individual designers working inside or outside the company to apply directly. The context also evolves very quickly, even while the design is being implemented, making it necessary to rethink the aims and contents several times while work is in progress, requiring specific types of know-how.

If we focus on human resource, it is evident that as territorial identities and those of local ethnic groups diminish due to mobility of people, companies tend to expand the sphere of involvement of skills on an all-round basis in various projects, and not just in geographical terms. Skills are no longer identified within a specific district, but in other industrial sites, companies, research centres and universities and without localized geographical limits.

10.1 Innovation Processes: The Solution, the Design, the Product

The acceleration of all processes requires effort and investment to such an extent that technological innovation can no longer be the outcome of the individual designer's intuition alone, but the result of a structured team. More and more often, the results of research appear to be almost simultaneous in different parts of the world: the competitive advantage for the company that invests in research lasts just

S. Palmieri (✉)
Dipartimento di Design, Politecnico di Milano, Via Durando 38A, 20158 Milano, Italy
e-mail: stefania.palmieri@polimi.it

M. Celi (ed.), *Advanced Design Cultures*, DOI 10.1007/978-3-319-08602-6_10

a few months and the innovation achieved must be ploughed into the product as quickly as possible.

Design becomes a sort of continuous flow, with the single product representing nothing but a photogram, and acquires the sense of a large-scale, overall strategic project that is identified with the thinking part of the company and becomes its value (Bartezzaghi et al. 1999).

Design is no longer a linear activity in which decisional and operational flows, controls and elaborations succeed each other, but on the contrary, is a complex process in which numerous subjects move in the same direction and interact in continuous transversal interactions, especially where the product is a sort of permanently developing abstraction which never stops being designed. The design process involves everyone in parallel and transversal ways, forcing the use of a shared language and synchronous timing, creating a sort of network in which design becomes a destination-meaning with respect to the continuous design process as often characterizes organizational systems. Continuous design is like a network in space: it is no longer a question of planning in terms of time, but also in a transversal direction. We no longer design in a single channel, with the customer on one side and the market on the other, but also horizontally, with the system co-suppliers to the right and left (Esposito 1996).

The organizational and cultural change that becomes necessary is considerable. Companies are forced to undertake aggregation inside and outside the territory; for the development of innovation the method that seems to respond the best to expectations and currently the most frequently used for the development of a design is teamwork. The group, with its multidisciplinary training and know-how, shares a purpose, has a common aim, works in close collaboration and shares the advantages of success. The team that allows for the contamination of skills works when its aims and methods are clear and shared, and when the whole group knows how to manage time, define and respect roles, procedures and rules.

It is here that the new role of design is embodied and emphasized: the idea of design cannot be separate from the idea of modernity, and designing means managing, understanding and expressing the reasons of our time. Due to its intrinsic characteristic of proceeding by interpreting modernity through organized as opposed to ephemeral creativity of events, design becomes a fundamental tool for organizing transformations: it becomes the main subject of innovation (Bettiol, Micelli 2005) .

The capacity to produce and manage designs forms the differential value of production systems, but also political systems where design is seen as the capacity and desire to organize resources, intelligence and culture, aimed towards an outlined strategic aim.

In this framework, *AdvanceDesign* is required to take on the role of stimulator of constant research and innovation, of promoter and mediator of various technologies, causing the evolution of methods compatibly with social and economic situations as well as industrial situations: *AdvanceDesign*, whether it is considered as experimentation on material structures or as a definition of new types of 'solutions' which precede possible products, has the mission of thinking of innovation before thinking of the design, before the project starts.

Design becomes an almost managerial approach, strengthened by its communicational and representational capacities, which become more and more important in industrial organizations and in the creation of future scenarios.

As already mentioned, innovation is deeply interfunctional and multidisciplinary, and is increasingly the fruit of interaction between different players in which the individual contribution is enhanced and the result goes beyond the mere sum of the capacities of individuals.

Every factor evolves and is subject to innovation at several levels. Product innovation is positioned at the intersection between the evolutionary paths of these factors (Bartezzaghi et al. 1999). It is a response to their evolution (the availability of new technologies for example) and a strengthening element (the change in models of use induced by a new product, for instance, with consequent effect on social behaviours). The success of a product innovation is linked to the capacity to integrate the factors of innovation and know how to pre-empt them.

For management, the fact that product and service innovation is positioned at the crossroads of the innovation of numerous other external factors implicates elements of outstanding complexity, due to the action of the innovative forces of the single elements and particularly their combination. Just like the strategic and organizational change, product innovation is one of the hardest challenges from the viewpoint of the innovative capacities of a company. This challenge is particularly evident today, where the turbulence of technologies, business models, markets and society has reached unprecedented levels.

For all these reasons, the growing demand for 'design' in increasingly complex fields imposes a critical rethink of the entire range of practices and logics that rotate around the design system and the economic-technical system as a whole.

The design sphere is required to take on new contents and offers new operating tools and methods to help prioritize the following issues:

- identification of the possible organizational–structural models of response;
- optimization of the impact on the product development process;
- contribution to the innovation process in a competitive way;
- definition of new future scenarios which can supply guidelines for designs and not necessarily for products (*Advance design*).

Then there is the attribution to design the job of governing and relating the many factors that trigger change, prioritizing the identification and orchestration of the numerous types of know-how and knowledge: hence, the need to know how to manage the interfunctional resources involved in the design team with effective leadership and make innovation, considered as a combination of creativities and know-how, a process which is systematic as opposed to episodic.

10.2 Enterprise Networks: The New Italian Industrial Model

Faced with these changes in the innovative contexts, it is possible to identify an authentic mission for advanced design, picking up on the changes taking place in the organizational structures of companies and in their links with the territories they occupy (Plechero and Rullani 2007). In recent years, the Italian industrial development model has undergone in-depth transformations, breaking free from the configurations of 'proximity' created in the past on the basis of the industrial districts (AIP 2008). These systems, which have been greatly appreciated and studied also by the international community, as the typical case of small industrial systems, have been overcome thanks to two precise drivers: market globalization and the growing content in terms of innovation and knowledge, disseminated in the various types of products/services offered to the markets.

The need to sustain the development of the products on the markets of various latitudes, with different cultural and consumption models, has made products and processes more complex, in order to adapt them to very different levels of performance and service. These multifunctional products often require several types of technology, consolidated in different sectors, which has determined their hybridization, with transmigrations between merchandising sectors.

Industrial organization has changed in response to this substantial economic need, forcing a break with territorial contexts characterized by one-way specializations. The migratory phenomena that have taken place over the past few decades have led to the partial loss of the sociocultural features of limited areas and led to the search for innovation. In short, we can say that the origin of the enterprise networks, on which the new Italian industrial model is based (Ricciardi 2003), meets the three needs deriving from these structural changes in the setting in which the enterprises operate:

- dematerialisation of value and hybridization of technologies;
- implementation of certain local specifics;
- expansion of the markets and of the competitors, and consequently the internationalization of business.

All of these phenomena can be attributed to globalization and the ensuing problem affects not only small businesses but also traditional large enterprises, which endure competition from productive systems characterized by considerable variability and ability to respond very quickly to change. In a certain sense, they indicate the mitigation of the power of manufacturers, offering easier access to information, and they generate interdependence of the markets and consumers.

Starting with the dematerialization of value, in the creation of the said value, which is typical of the production cycle, it changes the role of the phases that are capable 'of capturing' most of it, which are no longer (almost never) those of material transformation (manufacture).

This means that the intangible activities of the companies, which encompass consumer expectations and trends, from concept design to marketing and to applicative research and development, become particularly important.

However, the intangible value, which has become complex due to the multi-functionality and hybridization of technologies, depends on the ability to combine knowledge with the needs of the end user, and this also contributed to the structural changes mentioned.

Therefore, the value is concentrated among companies that control intangible production in terms of the ability to develop scenarios, design, innovation, trade-marks, communication, finance, commercial networks and services.

If we take a closer look at the phenomenon, we could say that its roots lie partially in the transformation of the design of parts of complex systems into smaller, dynamic organizations in support of big companies and partly in the acquisition of levels of greater knowledge via the creation of links with network systems, implemented by smaller companies.

The birth of the latest enterprise networks, which are currently characterizing the renovation of the Italian industrial system, coincides with the first 'supply' networks, the so-called 'short local networks' or 'local chains' (Cafaggi and Iamiceli 2007).

The current evolution, which is progressively adapting also in juridical and legislative terms, suggests, on one hand, an alternative industrial model to vertical Fordist integration, and, on the other, to the market autonomy of small enterprises, which characterized the country's first industrial development, under the slogan 'small is beautiful'.

The new model that is being consolidated can be defined as midway between the former two, in the sense that it strongly mitigates the features by integrating them vertically, defending small market autonomies, as long as they are linked together in networks, in order to take on the challenges dictated by the aforementioned technological and structural innovations.

The topography of the new industrial system is starting to change. Isolated small enterprises, incapable of creating a system for the search of innovative solutions and of taking advantage of relationships with research centres, are becoming increasingly rare, as are big companies that fail to expand their horizons and end up developing an autonomous, centralized know-how based exclusively on in-house specializations (Bonaccorsi and Granelli 2005).

The new generation of networks are becoming an option also for big hierarchical/functional organizations, when they become too rigid and slow compared to an environment which initially became turbulent and then complex, in which partnerships are often complementary or horizontal (Cafaggi and Iamiceli 2007).

Obviously, also in this new way of cooperating, which corresponds to the need to build networks, several types of networks and paths for their creation are being developed. They all combine reconciling the need to dismantle the structures of big companies with that to develop partnerships that are no longer restricted to the context of production phases but extend into what are known as 'soft' activities, including research and development and design, traditionally maintained within the individual, independent companies as an element of competitive differentiation.

This evolution can be defined as the creation of 'long-link' networks, which implicates specific business choices of connections which are studied before they are applied, making them more aware and cogent (AIP 2009). Without losing their independence, in this case, businesses have to decide to cooperate with one another, creating links of interdependence which, saving individual specializations and therefore certain levels of independence, pool some investments and the inherent risks by making a shared decision. We could say that the new frontier is a formula characterized by sharing and specialization.

This is a method that replaces casual and informal synergies of proximity with synergies of strategic aims and political desire to exploit the considerable investments required for participated or shared innovation.

The phenomenon regards companies of all sizes, although inevitably the creation of these networks requires cultural attitudes that lean towards sharing plans and an ability to combine different specializations, also in the field of the more traditional intangible activities.

We could say that broader and more articulate systems are being created, often with differentiations in roles due to the formation of an economic system that does not necessarily have to lose its local links but does not have to limit itself to them either, as they have become too restrictive for a globalized economy.

It is, therefore, possible to combine the survival of local links, which have become weaker, with the creation of relationships with no proximity restrictions. We are witnessing a sort of economic reorganization on two levels (nodes and networks), in which the old industrial districts can also survive, as long as they represent the nodes of a network to connect 'via long links' to non-local systems.

So far, we have examined the changes in the organization of macro business structures as the consequence of changes in the market scenario, both in terms of geographic dimensions and cultural, technological and social dimensions. The radical nature of these changes, which can be considered as authentic structural innovations, also has a strong impact on organization within companies. In particular, and obviously, structures destined to product and process innovation are strongly influenced because the products are changing and distribution in the market takes place together with service contents that are becoming more and more important. We must also consider the fact that increasingly frequent contacts and connections between companies and between universities and companies, companies and research centres at global level, make the way in which innovation takes place central to the company system. Innovation becomes a permanent way of building a business strategy, configuring innovation itself as one of the main levers for development of the company as a whole.

This results in the development of managerial content which is added to the traditional creative ability of designers and researchers.

We are all aware of the existence of relationships between creativity and innovation, in the sense that creativity and innovation concern the creative process and the application of new knowledge and have a real impact on the way of doing business on a much more widespread basis.

The term 'creativity' is often confused with that of 'innovation' and vice versa, but there are fundamental differences between the two terms. If we look at things from a broader angle, creativity goes beyond being a fundamental building block of innovation, which can be defined as the implementation of creative ideas.

Von Stamm (2003) for example, states that 'If innovation means turning an idea into something tangible, creativity emerges along with the idea in first place. Creativity is an essential part of innovation, it is the starting point'.

In the same way, Gurteen (1998) defines creativity as the 'generation of ideas' while innovation consists in transforming these ideas into actions via selection, improvement and innovation. Vicari (1998)shares this idea and states that creativity is input while innovation is output.

The shared opinion on the link between creativity and innovation stems from the conviction that internal business processes are similar to external ones. Given that quite often industrial sectors are characterized by an innovative change of type, which is the consequence of technological change thanks to a sort of 'technological-innovative chain', we tend to transfer the same succession inside companies and imagine a sequence of events caused by creativity, while innovation is the effect. In other words, creativity is transformed into innovation; according to this idea, creativity is the input and innovation is the output.

However, this concept of innovation is not suitable for industrial organizations.

It is better to state that creativity is not the source of innovation processes. It is not a sort of input for innovation, but the environment in which innovative processes can develop more easily.

The managerial content that characterizes innovative processes appears to be clear and, vice versa, the need also to implement managerial roles that are not specifically appointed to oversee research and development with creative contents.

Vicari (1998) also analyses the influence of creativity on management. In particular, managers have to have various skills depending on the level of complexity of the environment and of the competitiveness of the company in relation to its competitors. A common factor is that management, via an aware leadership, has to know how to motivate and direct the various and numerous contributions of people, holders of a creative potential regardless of the role they play in the business organization, from the managers to the operators.

Consequently, the decision-making process is carried forward with forms of collaboration between managers and staff, according to a participative management style. This said, in the economy of knowledge, where competition is based more and more on intangible factors, it is necessary to reward the search for and refinement of creative gifts, personal talent and the ability to apply it within work teams.

This has to be a company priority, desired and supported by management, along with its strategic aims. Incorporating the creative potential of staff in the organization of processes and strategies, in order to accomplish an innovation which is much more pregnant than technological and product innovation, which consists essentially in the ability to communicate and relate.

Naturally these strong innovations in the organization of intercompany activity and, therefore, particularly within the scope of network structures, require an appropriate formalization of coordination and control systems and of unitary governance tools.

The Italian legislator has recently approved laws which regulate the implementation of network contracts, particularly approving, for a certain type of network defined by experts as 'heavy' or 'associative' networks, the 'juridical subjectivity', i.e. the embodiment of the situation that is generated in a commercial contract, so that the existence of a stable and cohesive unitary subject, which can be accepted as the sole interlocutor in relations with third parties, is evident. Suppliers, customers, banks and public administration know that this type of network is a stable and cohesive unitary subject, which can be accepted as the sole interlocutor (AIP 2011).

This innovative legislation also simplifies the internationalization processes of the network contract which have been pursued this year, abroad and in Italy.

Looking at the situation from this point of view, those who do not work as part of a network or who fail to use the 'network contract' are not only more 'isolated' but also less efficient, possessing fewer skills and being, in the long term, less capable of proposing innovation for the market and, therefore, less competitive.

Furthermore, those who do not work as part of a network should not underestimate the self-exclusion from possible economic benefits deriving from industrial policies and incentives which, from now on, will definitely be studied, addressing them towards this type of support for the country's economic evolution.

In terms of the utilization of skills and know-how, this new organizational method will allow the use of *advance design* which, up to now, has been inaccessible to small businesses, thanks partly to the 'spreading' of the relative costs across several companies, several areas of merchandising and several products, thus enabling the Italian system to bridge the structural and cultural gap. This is practically obvious, because small businesses, even those that are most innovative, have never been able to consider—and have possibly never even thought about—the possibility of pursuing an activity which is not applicative.

At this point in time, we cannot exclude thoughts of developing professions connected to the creative and design sphere, especially considering that, in the author's opinion, organization to networks will probably develop the activity of professionals as possessors of skills which cannot be incorporated into stable business structures and which are more easily available through independent consulting relationships.

10.3 Technology Parks

In the framework outlined, Technology Parks are taking on an important role and are undergoing considerable revitalisation, being configured more and more often as aggregations of different public or private stakeholders, to create innovation through the collaboration among companies.

They are centres for the development of research, generated by initiatives of big companies, small businesses or public entities.

The aim of a multidisciplinary scientific technology park is not merely to create radical innovation 'from nothing', but also to allow a technological exchange that favours the imposition of a process, an application or a product also in different sectorial spheres than that for which it was designed.

Three cases have been identified which, for different reasons, supply a picture of the trends in progress, as follows:

- Campus Innovazione in Chieti, for the automotive sector, a publicly funded technology park set up by Fiat and Honda
- Kilometro Rosso, created around the Brembo company, which currently works with private funding and serves industrial development
- Ecsa, which operates in the aerospace sector and involves SMEs, Universities and some Public Centres.

Case 7: Campus Innovazione Automotive

> By the end of 2017, Abruzzo will have an automotive campus, a strategic structural element which looks beyond the boundaries, thanks to the four laboratories for research, training and new technologies and a test track for carrying out direct tests, not in competition, but qualified and technical ….An infrastructure aimed entirely at innovation and research, which will make the community based in Abruzzo stronger in its approach to the challenges of globalisation (Gianni Chiodi, President of the Regional Council of Abruzzo, August 2013)

A significant case of technological innovation and development of business culture is the Campus Innovazione Automotive in Abruzzo. It has been chosen in one of the key sectors of the country's economy, the automotive field.

Why this choice?

1. First of all, because the automotive sector is one of the busiest in the country at the moment and represents, along with the fashion business, food farming and furnishing one of the 'four A's'—Automobili, Abbigliamento, Agroalimentare and Arredamento (automotive, fashion, agroalimentary and furniture)—of the well-known Made-in-Italy (Cietta 2008).
2. Because the automotive sector is characterized by extensive space for advanced design, seen as experimentation of material structures and as definition of new types of 'solutions' which precede possible products, requiring considerable reliability and therefore superior quality standards, also in terms of environmental eco-compatibility. These are all things that can be found in the case in question, also as a dimension prior to the applicative phases.
3. The third interesting element is related to the stakeholders involved in the project. They allow a project with a strong territorial value to achieve an international dimension, assigning to the Automotive Campus a globalized value.

The Automotive Sector

In Italy, the figures for the sector (car industry, sub-supply, components, engineering, design and the distribution network) are:

- 2,500 companies, many of which operate in the component sector, with 60 % in the north;
- 165 billion euros in turnover, representing 11.4 % of the domestic GDP and 30 % of the manufacturing industry;
- 2 billion euros per annum of investments in research and development which represent 22 % of the total expenditure in research and development of private Italian companies;
- 81 billion euros of tax revenue;
- 400, 000 operators.

In Abruzzo the mechanical-automotive and electronic sectors represent about 31 % of the industrial GDP; this value rose in 2007 and in 2008 but it is dropped in 2009–2010.

The main players in the region are represented by Sevel, Fiat, Honda, Denso, Pilkington, Dayco, Honeywell, Pierburg, Tyco, Imm, Tecnomatic, etc.

The automotive sector is very demanding in terms of continuous technological innovation, an essential requisite in order to undertake the challenges posed by the international markets. The transportation sector in particular, records a rather high (0.55) national Sectorial Innovation Index (Innovazione Settoriale—ISI), clearly higher than the average for the manufacturing sector (47). Moreover, the metal-mechanic sector in Abruzzo shows an increasing need of innovation related to the manufacturing sector. This is particularly marked in the areas of production, design and marketing.

So a critical issue is the basic lack of technology skill in the Abruzzo's companies. It still hinders a rapid growth especially in sectors such as the metal-mechanic and automotive sectors. It is due to the unwillingness of the SME to invest resources in studies, experiments and prototype designs, etc.

These key points lead to the creation of a specific Automotive programme oriented towards the automotive and mechanics sectors with the aim to allow technological and systemic experimentation. It represents an initiative capable of developing the dissemination of know-how and innovation.

For the envisaged activities and especially for the identified sectors, the project has national relevance and it can be a development element for the whole Italian productive system, even if it is focused in Abruzzo and in the Chieti's province.

The Campus

The Campus dell'Innovazione concept comes from the synergy among productive realities (Big Enterprises and SMEs), local authorities, schools and territory. It is a pole of attraction for research, development and advanced design to merge theoretical and practical aspects.

The Campus has been opened as the evolution of a Consortium, the CISI. The CISI grew up spontaneously around the Honda company in 1992, and was created by 17 companies specializing in the production of components and services for the motorbike industry, with the aim of creating a tailor-made supply chain. The Consortium worked for about 15 years, but without relevant results in terms of innovation. Since 2007, thanks to the intervention of Fiat and other institutional interlocutors, the initial group of 17 companies become a more structured Consortium focused on the semifinished components' production for the automotive field.
So the production chain consists in cross-sectorial elements that create a sort of 'knowledge network'. Thanks to the partnership with the University of Bologna and with the Confindustria in Siena, there is the development of product and process know-how and the improvement of innovation skills.
The campus dell'Innovazione Automotive e Metalmeccanico is situated in the Sangro-Aventino's. Its mission requires the participation of Local Authorities (Provincia di Chieti and other Entities), the Enterprise System [1] and the University. The activity of the Campus aims to bring innovation to all levels of the production chain, translating into a dissemination system. Innovation is seen as a process. In particular, the design process helps to divide the creative path into separate phases: from the definition of the strategy, to the analysis of the consumption scenarios and experimentation of concepts, arriving to the development of a product (Fig. 10.1).
Considering that is not possible to have any kind of innovation without the evolution of knowledge, human resources and cognitive and metacognitive tools, the Campus is realized as a centre for developing and updating skills to create added value for the whole production's segment (Fig. 10.2).

General Aim

The Campus' general aim is to develop an automotive system related to lights for commercial and professional vehicles (two/four wheel transport for people and goods), capable of strengthening and consolidating the production chain to improve competition and favour the rooting of multinational companies in terms of:

- system innovation
- product and process innovation
- flexibility of products and their manufacturing processes
- product quality

The programme aims to increase the skills and improve the know-how dissemination in the automotive and metal-mechanic sectors to achieve development:

[1] With reference to the automotive sector, we wish to indicate FIAT-SEVEL (with the Fiat Research Centre) and its chain, HONDA and CISI Group, Denso, Pilkington, Dayco, IMM, TECNOMATIC, ASTER (HI-MECH district), etc.

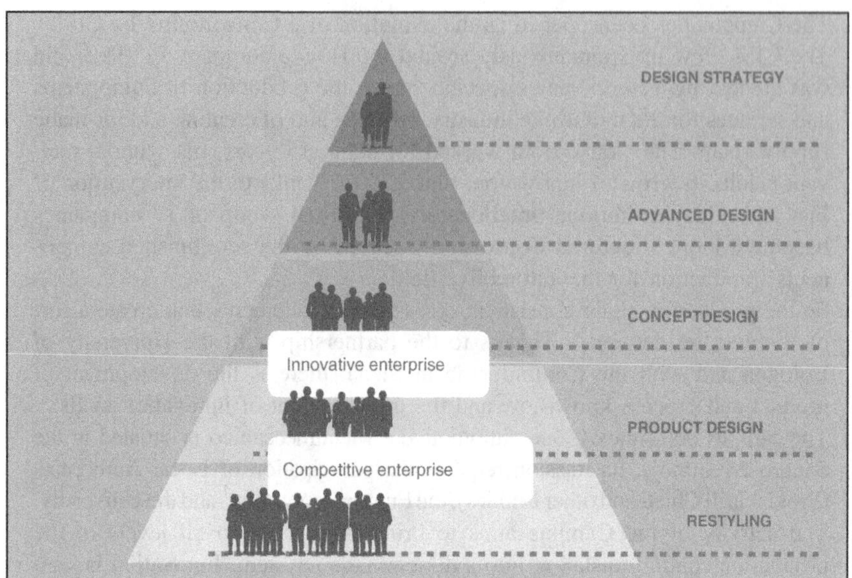

Fig. 10.1 Innovation as a process: the design process (*source* automotive production chain integrated programme: summary—2007)

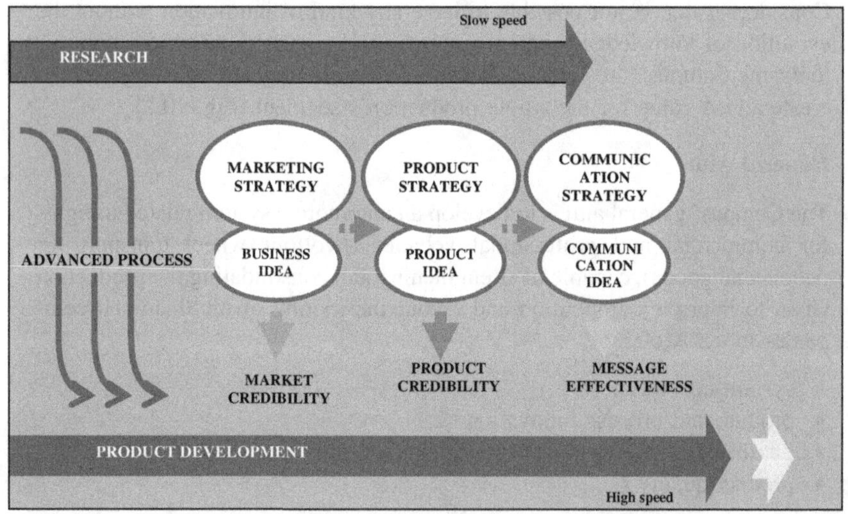

Fig. 10.2 Innovation as a process: advanced design provides vital input to various company departments to guarantee the consistency of their work with the initial indications of the project. (*source* automotive production chain integrated programme: summary—2007)

- integrating the research and development of innovative solutions, training and business culture;
- favouring the interaction between the players of the production chain and the synergy between different industrial sectors.

Case 8: Kilometro Rosso

Kilometro Rosso is a scientific technology park built next to the A4 motorway in Stezzano, near Bergamo. Set up by Brembo, world leader in the production of braking systems, Kilometro Rosso represents a multidisciplinary research and development centre oriented to the district's knowledge growth, innovation and high technologies. Kilometro Rosso is not the Brembo research and development department. The company needed to move its research and development laboratories away from the production centres. Indeed, in the previous configuration the centre work was constantly disturbed by the environment and the main effort was focused on the products' quality and defects and not on the radical innovation.
The strong idea of this installation is that it would not only be destined to the research and development of Brembo, but would become an authentic scientific technology park, also open to other research experiences and needs which could come from the outside.
Behind this strategy there is the conviction that 'innovation cannot be achieved alone', and the collaboration is necessary.
Consequently, the decision was to develop research activities in different disciplinary sectors which could contaminate each other.
In the Kilometro Rosso spaces there are a lot of companies. Some of them are not related to the automotive sector, such as Mario Negri Institute and Italcementi. Indeed, the centre is the node of a broader network, with different research focus. In this sense, it is important to note that Kilometro Rosso is in partnership with important international institutions, such as the MIT.
In relation to the sub-networks, there are a lot of interesting activities and synergies, for example, the creation of Intellimech, a consortium of 26 high-tech companies dedicated to interdisciplinary research within the scope of mechatronics, which comprises advanced electronic design, IT design and the design of ICT systems.
From 2010, Kilometro Rosso have about 20 partners and 1,500 research and development operators, (there are currently about 700). The aim is to reach 50–60 partners and 3,000 operators.
The leadership of the park is linked to Brembo, which conceived, designed and built Kilometro Rosso. However, it is managed by 'Kilometro Rosso srl', a management company with the aim of developing the systems of internal and external relations of the scientific and technology park.

Case 9: GEIE ECSA, a Cross-Fertilization Network

The ECSA (European Centre for Space Applications) is a network of multidisciplinary, international scientific excellence, comprising European small and medium enterprises (SME), universities and research entities. The ECSA is an EEIG (European Economic Interest Grouping) and is qualified as a Research Centre by Regione Lombardia.

The aim of the ECSA is to pursue advanced research projects in the spatial sector and promote collaboration between industry and university.

In 2004 a small group of SMEs operating in the ICT, albeit in different industrial sectors but with particular reference to spatial applications, decided to create a collaboration network to achieve a dimension capable of competing on markets undergoing rapid internationalization. The strategic aim for its formation was to create a new organization which could facilitate small enterprises' participation in tenders promoted by the European Union, by the Italian Ministries and also by the Regional Councils (with particular reference to Regione Lombardia, where most of the partner companies are located) to create innovation projects in keeping with the development aim.

The EEIG was set up in 2005, at the 'Polo Scientifico e Tecnologico Lombardo' in Busto Arsizio (Varese), to manage and promote the 'European Centre for Space Applications (ECSA)', centre for applied research and technologies' transfer, which intends to operate with SMEs and Universities/research Centres in the sectors of monitoring and assessment of environmental data, including that procured from satellites, renewable energies, space, robotics, ICT, transport and logistics, etc.

There are currently fifteen partner companies, five research centres and seven universities.

The headquarters has recently moved to Benevento to encourage the evolution of the grouping towards a market logic.

Each partner company independently develops its core business: the EEIG ECSA does not concern itself with this type of small business, proportionate to the small dimensions of the partner companies, but has to promote and manage innovative initiatives of considerable breadth, which the single partner company doesn't always have the possibility to identify and therefore to promote.

The network cannot function if all the partners have to collaborate with each other, but it functions properly and develops if it is capable of identifying, designing, promoting and managing big international projects in which different skills are necessary on the technical front and also in terms of promotion, organization, management, etc.

References

AIP: Reti d'impresa oltre i distretti. Nuove forme di organizzazione produttiva, di coordinamento e di assetto giuridico. Il Sole 24 Ore Ed., Milan (2008)

AIP: Fare reti d'impresa. Dai nodi distrettuali alle maglie lunghe: una nuova dimensione per competere. Il Sole 24 Ore Ed., Milan (2009)

AIP: Reti d'impresa: profili giuridici, finanziamento e rating. Il Sole 24 Ore Ed., Milan (2011)

Bartezzaghi, E., Spina, G., Verganti, R.: Organizzare le PMI per la crescita. Il Sole 24Ore Ed., Milan (1999)

Bettiol, M., Micelli, M.: Competitività dei distretti e design: rinnovare le basi della creatività. In: Bettiol, M., Micelli, S (eds) Design e creatività nel made in Italy. Proposte per i distretti industriali, pp. 105–132. Bruno Mondadori, Milan (2005)

Bonaccorsi, A., Granelli, A.: L'intelligenza s'industria. Creatività e innovazione per un nuovo modello di sviluppo. Il Mulino, Bologna (2005)

Cafaggi, F., Iamiceli, P. (eds): Reti di imprese tra crescita e innovazione organizzativa. Il Mulino, Bologna (2007)

Cietta, E.: La rivoluzione del fast fashion. Franco Angeli, Milan (2008)

Esposito, E.: Le imprese ad alta tecnologia. Il caso dell'Industria Aeronautica. CUEN, Naples (1996)

Palmieri, S.: Il consumatore che innova. Il Sole24ore Ed., Milan (2004)

Plechero, M., Rullani, E.: Innovare. Re-inventare il made in Italy. Egea, Milan (2007)

Ricciardi, A.: Le reti di imprese. Vantaggi competitivi e pianificazione strategica. Franco Angeli, Milan (2003)

Von Stamm, B.: Managing innovation, design and creativity. Wiley, London (2003)

Gurteen, D.: Knowledge, creativity and innovation. J. Knowl. Manage. **2**(1), 5–13 (1998)

Vicari, S.: La creatività dell'impresa. Tra caso e necessità. Etas, (1998)

Chapter 11
AdvanceDesign: A Renewed Relationship Between Design and Science for the Future

Marinella Ferrara

> *Any time one or more things are consciously put together in a way that they can accomplish something better than they could have accomplished individually, this is an act of design.*
>
> Charles Eames

In each phase of the industrial revolution, changes in paradigms faces designers with new questions and challenges.

The third phase of the industrial revolution we are living these days is characterized by the changes brought by digital technologies, their integration with sciences, and how rapidly technological and scientific discoveries turn into applications, affecting our ways of living, puts us a question: what does design mean in a society going through continuous change?

Nowadays, the topic of the sense of design feels uneasy because we have lost every hope of acquiring to a rationality able to come to complete description and comprehension of reality. The complexity of technical systems is growing, as well as the tragic consequences for the environment and our health by short-sighted usage of technology combined with uncontrolled industrial development.

The loss of certainty has already arisen in many fields of contemporary thought, just think of the fall of the 'linearity of science', from Kuhn to Bruno Latour (1987). The latter has explored the mechanisms of reception of discoveries (laboratory activities, the role of scientific literature, institutional context of science) and has brought to light the need of considering the work of scientists and technologists just like any other social activity, in order to improve actual understanding of scientific activity. This issue, as well as many others, brought us to turn the idea of a deterministic and predictive science into that of a qualitative and hermeneutic one. Giving up complete domain on reality brought to favour weak and reflective rationale patterns, based on complexity management, process reversibility and social value of knowledge. Problem description and solution negotiation moves

M. Ferrara (✉)
Dipartimento di Design, Politecnico di Milano, Via Durando 38A, 20158 Milano, Italy
e-mail: marinella.ferrara@polimi.it

M. Celi (ed.), *Advanced Design Cultures*, DOI 10.1007/978-3-319-08602-6_11

from past institutional contexts—government, industry and university—into a space where 'science meets public and public talks to science' (Nowotny 1994).

Since there is no doubt that our future will be more and more marked by development of robotics, nanotechnologies and digital logic, which are actually changing the ways of designing, debate on innovation and its paths is a central theme, related to benefits it actually brings, apart from those needed by companies to compete on the global markets (Pasca 2009).

In this situation, which is the role that design culture plays today? On which side of the debate about science and technology do designers place themselves? In favour of the sensible use of technoscientific discoveries, or against them? In favour of a critical reflection on innovation, wondering about its objectives, or the 'status quo'?

The thought and discussion about this theme is very important in that today's reaction to some kind of *violence* in technocratic rationality has become more radical (Maldonado 1992) in the respect of return to lifestyles which, as Pasca (2009) affirms, are inspired by communitarian nostalgia and banning of science by ecoterrorists. We do perceive a wish for mediaeval and communitarian lifestyles as John Ruskin hoped in the nienteenth century, when he raged against trains and the evolution of technology.

Can design itself be a discussion platform about future choices and scenarios?

My answer is yes, but which tools and methods must it adopt in a condition of permanent change?

Some features which make design research ready for contemporary challenges are: ability of comparison and sharing of knowledge and among different knowledge fields, openness to the world, accessibility, non-structured thought, as well as awareness of weak signals and hidden stimuli.

Furthermore, debate between technoscience and *humanities* is rich soil for critical innovation. Going towards possible, attractive, desirable or deniable *futures* (after definitions adopted by futurologist Start Candy) with contributions by technosciences on one side and disciplines like sociology, anthropology and philosophy on the other, is an exercise which can stimulate new thoughts and design approaches, as well as bring to attention an *advanced* notional dimension, which is not immediately aimed at mere competitive advantage for companies, but can give shape to new priorities in comprehension and usage of today's technological potential.

The future of mankind has gained much interest for many academic institutions, like the Future of Humanity Institute at the Oxford University, the Center for Responsible Nanotechnology, the Institute for Ethics and Emerging Technology, the Methuselah Foundation Program and the Singularity Summit. The comparison and cross-contamination of the ideas these institutions promote can contribute to the creation of a culture capable of supporting scientific and technological innovation, beyond the tracks of consolidated knowledge, towards the horizon of experimentation.

Research laboratories like MIT and companies like Philips, Nokia and Samsung, which are practicing Advanced Design, even if often not expressly, are competing on the design themes arising in relation to the scientific discoveries in computer science, physics, biology, medicine, the nano- and biotechnologies and neurosciences, stimulating the creative component of design in search of new opportunities for the future.

The projects in this centres are the objective of turning previsions into scenarios in alternative to the *status quo,* in which the increase of benefits is equivalent to the reduction of the negative effects of industrial development through consolidated processes.

Many designers express the need for a rethinking of the consolidated process spreading from project to industrial product to launch on the market, and for some time already we can see a mix-up of roles, practices and timelines. We can perceive the need to escape from the industrial and market practices, to push the innovative process towards alternate visions and new sorts of knowledge, to ban the stress for immediate result, which defines the temporality in professional practice without adding value to investments capable of creating structural qualities within the mid- and long term, and to pursue an experimental approach with a reflective and critical vision of the world and our own operational method.

Also at the Politecnico di Milano, the *AdvanceDesign* research contributes in pursuing an intent of reflexive innovation to underline the goal to research alternate visions for the future, we wonder how we.

The *AdvanceDesign*, synonym of exploration and experimentation, can start a renewed and intense relationship with science, a rich environment for really advanced processes, with the aim at contributing with new ideas to dramatic innovations, which require experimental work methodologies and deep and broad knowledge. To explore transformation processes taking place now, to experiment emerging technologies, to evaluate the potential of new materials and to propose design solutions to be turned into new generation of products, drives us to use imagination and to the construction of a better world.

The relationship between design and technosciences is well promising to that purpose. It will contribute improving capability to design the future.

11.1 Design Research Exploring the Future

In 2009, the *Design and the Elastic Mind* exhibition hold at the NY MoMA focused media on the theme of the upcoming relationship between design and science.

Curator Paola Antonelli has selected and presented researches which relate design competences with other disciplines (physics, mathematics, computer science, management engineering, chemistry and bioethics) in order to highlight the role of design in facing fast rhythm of changes characterizing contemporary life.

This exhibition proposes a new definition of design research as the ability to grasp the meaning of changes in science, technology and social habits, and turn it into product and service concepts.

In this occasion Paola Antonelli has pointed out: 'Designers belong to a new culture, where experimentation is stimulated by what happens everyday, and *is open to cooperation with colleagues and other specialists [...] Consideration for the role of human dimension in life is central to the emerging dialogue between design and science.*' The words confirm the designers' ability of bringing back technoscientific discoveries to the physical and mental dimension of common man as a fundamental issue for their comprehension and usability. And more, this process itself often contributes to increasing knowledge, the more revolutionary are scientific discoveries, the more important is the mediation of design, which makes their introduction in everyday life easier, avoiding society to be passively overwhelmed by change or feel threatened by it. Just like when we have to match to the human scale, i.e. to concepts easily understood by man, the infinitely small or large dimensions towards which science and technology have expanded in the last decades. (As in the case of infinitely small or large discoveries towards which science has expanded in the last decade, which must be translated into concepts easily understood by common people).

Thanks to its broad and complex capability of exploring and foreseeing, design is capable of operating the mediation between science and everyday life at different levels: at the instrumental one, which allows to 'shape' what is invisible to human eyes, at experimental and planning one, which considers the daily use of technical and scientific discoveries, at the scenario one, which affects the vision of the world, on the ability to direct innovations.

11.2 *AdvanceDesign* as a Tool for the Communication of Science

It is about the instrumental level of mediation of science when design contributes to communication of scientific discoveries: as an example, the way nanotechnologies are visualized and explained is central to their correct comprehension and use, in order to promote constructive dialogue between scientists and colleagues in the same or complementary disciplines, and also to allow wide, non-expert audiences understand scientific concepts.

Design can contribute to visualization of nanoscaled processes in order to enable the scientists themselves through a constructive communication with colleagues and complementary disciplines, to spread the communication to a wider audience and let the end-user understand the characteristics of a product conceived in nanoscale.[1]

In the design history, some designers have contributed to this kind of mediation: in 1977, Charles and Ray Eames, with their Powers of Ten video, an exciting journey through the dimensions of universe and quarks, which has become a classic in scientific popularization, could foresee and communicate the widening of human

[1] Goodsell (2009) has underlined how such visualization is not free from ambiguities.

perception due to technoscientific discoveries in the infinitely small and large scales.

Design working on visualization of complex scientific concepts, making use if various analysis, synthesis and rendering processes, like those described by Kemp (1999), plays an important part: thanks to its contribution, some significant episodes of science have been decided thanks to the information contained in images, thus realizing an actual growth of scientific knowledge.

On this side of research, expertise in 'information design' is very interesting. This term stands for 'comprehension design' after definition by Richard Saul Wurman, the inventor of TED format. Information design competences are aimed at creating perceptive frames, which can turn complex data into information giving an easy, clear, accurate message.

Currently, many researches are addressed to creating perceptive devices and display tools which help understanding information from scientific reports and the invisible natural systems surrounding us, like air and water properties, or artificial systems like flows of energy, of materials, of polluting agents, as Bruce Sterling suggests in his SPIME (2005).

The unveiling of what is hidden behind every action and daily choice, contributes to the increasing of understanding, and hence awareness, an essential requirement for cultural change. Like Nathan Shedroff, a pioneer in experience design, affirms, comprehension is the process of transforming information into wisdom. This process is simplified by memorable information, i.e. completed by content organization and narrative through words, images and graphics.

11.3 *AdvanceDesign* as an Explorative Practice

Another level of the design mediation is the experimental and planning one, in which we study applications of technology, which can open new opportunities in terms of solutions aimed at improving life quality, beyond existing objects and environments.

Scientific discoveries may inspire new usage ideas, and applications unpredicted by scientists, whose notional thought often excludes ordinary human life matters, like relationships, illnesses, death or sex, which instead stimulate the designer's attention.

Advanced Design is therefore an explorative practice, before project development. It is a kind of activity synthesizing the cognitive value of experimentation, independently of a fixed final application goal.

In relation to this, we can mention projects by James Auger and Jimmy Loizeau from Human Connectedness Group at Media Lab Europe in Dublin.[2] In this centre,

[2] Independent research institute and excellence centre in digital technologies. Founded in 2000 by the cooperation between the Irish government and MIT in Boston.

Fig. 11.1 James Auger and Jimmy Loizeau, Iso-Phone project

they reflect on possible implications of ICT usage on individuals, society and objects.

Through an approach combining engineering, product design and art, we can predict new application scenarios of technologies, we can investigate through design experimentation in their quality dimensions for human behaviours.

We can consider, risks of social isolation, alienation from reality, stress, due to the acceleration of information flows and to the new dimensions of real-time connectivity as well as virtualization, which characterize present forms of communication. These researches amaze us with technological prosthesis like the cell-phone-in-a-tooth or the *Iso-Phone*, an helmet to be worn underwater, which isolates from any external stimulation, bringing our attention back on the speech component of phone conversations. They propose new scenarios in social interaction where deepness, conviviality and quality of human relationships are not cancelled, but instead enhanced and supported by technologies (Fig. 11.1).

Critical Design by Anthony Dunne, director of the Interaction Design Department at the Royal College of Art in London, and Fiona Raby, is active in researching worldwide social, environmental, cultural and ethical implication. Their design is a means for stimulating reflection and for encouraging debate among designers, industry and public administration about social, cultural and ethics implications of emerging technologies. They realized unusual objects which do not seem to have any practical use, at least none of those we usually think of, devices which embellish our homes, in order to explore aesthetical, perceptive and functional capabilities of electronic products. Those researchers think design as a means

Fig. 11.2 Anthony Dunne and Fiona Raby, Placebo Project

of going over preconceptions, stereotypes, *clichés* about products of daily use. Their *Placebo Project* (Fig. 11.2) is composed of eight prototypes which react to magnetic fields in different ways, like *Parasite Light*, which lights in the presence of radio waves and adjusts its brightness *sucking* electromagnetic field waves, like parasites do. Or like the *Electro Draft Excluder*, a shield against radiations, or the *Compass table*, with a surface full of needles which oscillate in every direction when a cell phone or notebook computer is laid on it, and more, the *Nipple Chair*: when crossed by an electromagnetic field, it will warn whoever sits on it through radiations spread by two vibrating needles in the back.[3] Their research is about relationships and behaviour of people who live more or less consciously in contact with electromagnetic fields, surrounded by electronic devices. Designers have assigned eight prototypes to as many volunteers in order to investigate their reactions while objects work. Those prototypes, simple in their shape and material, reacting to what invisibly happens surrounding environment, and initially amaze their user and make them aware of what surrounds himself. Reactions to the energy flows revealing 'the secret life of electronic objects', demonstrating that, in everyday life, objects can perform far beyond the producer's imagination.

---

[3] Tony Dunne and Fiona Raby, professors at Royal College of Art's Design Interactions Department, said: 'Electronic objects not only are 'pleasant'. Being able to imagine them in their subconscious aspect rather than in their shape gives us new interpretations. For example making objects interact with electromagnetic fields, we integrate them with space and man, who lacks those perceptions. This is their peculiarity' (Gorman 2009).

Critical Design by Anthony Dunne and Fiona Raby is a design experimentation method as a preliminary stage of the design itself. The goal of the research is discovering the possible, desirable or objectionable relationship between the unknown object and the end-user. Experimentations deals with pointing out potential brief and concept, which can address design research from problems to feasible solutions.

Therefore, as now, with the *Placebo Project*, they investigate through prototyping the behaviour and reactions of users, the next step will possibly be to provide an adequate protection from electromagnetic waves.

Not immediately directed to the application of new technologies to objects, but only to evaluate the opportunity of their deployment, this method is able to drive application results up to definition of new research goals and area of interest for designers.

11.4 Drive Innovation Through Advanced Scenarios

Advanced Design practices involves the participation of expertise from different disciplines (chemistry, engineering, biology, computer science, etc.) building connections among various scientific theories, and opens design to opportunities of disruptive innovation.

Design, acting as a mediator, plays as knowledge tool and, at the same time, is a transformation factor in regard to the new social movements. The designer thus plays the role of a 'technical professional intellectual' (Riccini 2009).

Thanks to design mediation, scientific research enters the dimension of 'productive making' and works in the most urgent problems in the world. It leaves the neutrality and disregard position about the consequences of its work, and places itself historically and socially, to contextualize its results.

It thus happens that new technology potentials are directed towards the principles of environmental sustainability, security issues, health and sustainable economic development.

In present days, the thought about high technologies and the researches named Earth Care Design promote the reorientation of technology towards the principles of Nature and bioethic, through the biomimetic scenario.[4]

In the biomimetic vision of design nature is considered as a 'model, measure and guide' to the design of technical artefacts. Biological qualities can be transferred like performances to the design of objects and architectures. Indeed, design solutions to draw inspiration from (Benyus 2002) can be found in the wide number of

[4] Biomimetics is a discipline derived from bionics very popular in the 1960s. Compared to bionics, which appeared to mimic organic structures to produce more efficient products, biomimetic has a wider vision. It does not merely reproduce automatically the forms and structures of organisms, but proposes to build on the strategies and logic that underlie the evolutionary success of biological systems.

intelligent, sensitive and sustainable structures found in nature (following appropriate strategies, self-learning and self-organizing logic, using less matter and energy and are more efficient than traditional productive systems). So, design problems are confronted to biological phenomena. Analogies create new solutions, and show new design paths as a translation of natural logics.

Among the various approaches theorized by multidisciplinary groups, 'Cradle to Cradle' is published by the American architect McDonough and the German chemist Braungart (2002). Their strategy, presented in their homonymous book, has the goal of overtaking the usual environmentalist approach—reduce, reuse and recycle—by adopting some working models found in nature. The theory suggests that produced 'waist' could be useful (nourishing) components for the growth of object systems, like biological ones, based on an ecosystem logic which realizes the maximum level of technical metabolism, preserving variety and differences more than homogeneousness. The three principles of 'Cradle to Cradle' contribute with a major change in the way of building the world. First principle: convert waist into nourishing matter and food, and remove the concept of waist. Second: exploit the energy from the sun and environmental phenomena. Third: celebrate and promote biodiversity.

Such ideas can help industry balance the interest on technology in favour of the 'complex organism plus its environment', to use an expression by Bateson (1972).

Some researches within the Interaction design program of the Royal Collage of Arts in London explore the bioethics territory. Particularly, the immortal and Life Support di Revital Cohen, a series of organ replacement machines connected in a semi-biological circuit. Revital Cohen develops critical objects and provocative scenarios exploring the juxtaposition of the natural with the artificial. Her work spans across various media and includes collaborations with scientists, animal breeders and medical consultants.

Life support envisions domestic animals transformed into medical devices that claim the live of their owners. For example a retired greyhound could be retrained and used to help a patient dependent on mechanical respiration (Fig. 11.3).

Could animals be transformed into medical devices? Could a transgenic animal function as a whole mechanism and not simply supply the parts? Could humans become parasites and live off another organism's bodily functions? These are some of the questions the project places to itself.

Another area of interdisciplinary reflection is that regarding sustainability of development, named *Social design* by Margolin and Margolin (2002), who defines its goals as:

> Social Design is that contributing to social wellness […]. One of the goals of social design is to reach those who presently do not benefit from design. Another is to produce goods and services which avoid the negative effects of the most of what we presently produce.

Visionary researchers like Neil Gernshenfeld follow this track. He is the director of the Center for Bits and Atoms at the MIT, and has worked for years to study the relationship between information and physical properties through which it reveals. After having researched for long time, other tight relationship between these two

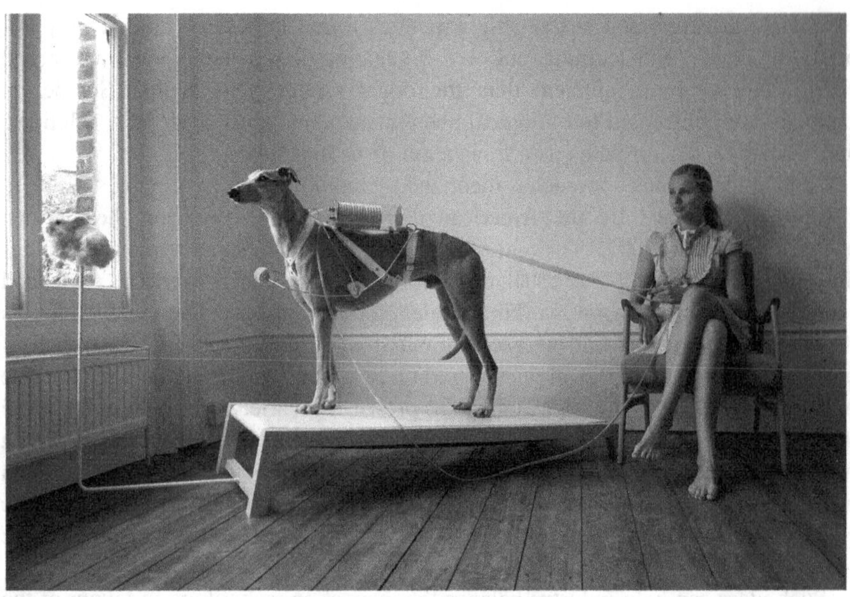

Fig. 11.3 Revital Cohen, Life Support

subjects, which stimulate us with the most interesting and difficult questions for the near future, with the first 50 FabLabs around the world, he could trigger a productive revolution starting from poor countries like Afghanistan or Ghana, or in emerging countries like India. His project responds to a strategic vision: to install into villages and districts in the world small-scale laboratories, equipped with advanced technology and easy to use, network-connected machines, to provide local populations access to the tools for rapid manufacturing, thus stimulating the establishment of local scale productive activities, which could produce 'almost anything'. FabLab's promise is to offer independent and autonomous chances of prosperity to the poor people of the world, without relying on past industrialization logic.

The social relevance of this project lies in the possibility to offer poor and uncomfortable populations education tools for competence development and independence of industry in manufacturing. This could favour a revolution in the production world through social creativity stimulated by the access of a new meaning of inventing. The impact could be epochal: mass products, chosen for lack of alternatives, would not exist any more, but—almost—anyone could create unique objects by itself, and for his own needs. This would allow the configuration of a new, advanced dimension of handcrafting activity, advanced, i.e. capable of following transformations in crafts, and to open to new forms of expression, to new techniques, to creativity and innovation.

11.5 Advanced Material Design

In the framework of the new relationship between design and science, as just depicted, a particular field of design research development is taking place: Advanced Material Design. This puts together different disciplines: competences in material science (mainly based on chemistry and physics, i.e. knowledge of solid bodies, properties), and in material engineering (where attention is mainly directed to engineering aspects, from production processes to mechanical performance), are integrated with design-specific ones, with great attention to perceptive-sensorial qualities, to functionality and performance, and to the scenarios of environmental sustainability.

Since long-time material research has not been an exclusive matter for chemists, chemical engineers or material scientists: designers have been offering a substantial contribution to material innovation, building products in which the qualities of materials, giving an aesthetical and functional value, have brought the need for new languages. Through experimentation of shapes and structures, and using materials available each time, designers have increased their knowledge of the technological characteristics and have contributed to the evolution of materials, often stimulating real process innovations, and to the research for new ways of using materials and the definition of their actual expressiveness or new designed qualities.

In the 1970s, the Italian 'design primario' opened a new and acknowledged field in design for research within design discipline: material design. It happens where materials acquires its bundle of chromatic, acoustical, visual and surface properties, to give the material its own identity. Shifting attention towards the soft quality of matter (colour, light, sound, smell, texture), the design culture balances the hard qualities, those due to formal structural composition, and focuses on sensations, on physicality as an experience, on the communication values of materials.

This opening to the qualities of materials in research is contemporary to the development of composite materials, which are paradigmatic of material design at the macroscopic level: by combining several materials together, it is possible to obtain a new material with better properties than the single components and surprising properties if compared to usual materials (resistant and lightweight like carbon fiber compounds, or rigid and flexible at the same time, like woods coupled to plastic).

In 1995, the MoMA, holding the 'Mutant Materials in Contemporary Design' exhibition, gives attention to the evolution of materials driven by design, exhibiting the researches which realize perceptive displacement in relation to the consolidated idea of materials: ceramics harder than metal, woods as soft as tissues, sinuously curved glasses, etc.).

In the last 25 years, a wave of scientific discoveries have made our relationship with matter evolve quickly. The electronic Scanning Tunnel Microscope (STM) has favoured our comprehension and control on materials at the nanoscale.

In this way, today technology enables design to push forward its horizon. New materials are built atom after atom, molecule after molecule, with simple or complex characteristics, as is the case of smart materials, with their own behaviour. Thanks to the improvement of computerized analysis methods and to characterization of materials and of instant production methods (deriving from rapid prototyping), it is now possible to design material structures capable of working in a particular way to respond to peculiar requirements in different application fields.

The increase of technoscientific knowledge results into higher project complexity. Performance becomes a priority, just like the quality of interaction of individuals with their material environment (made of ambient and objects), and this has become one of the goals of design research. Design thus goes into detail about the comprehension of sensorial and perceptive processes both in what concerns the physical-sensorial and psycho-perceptive components of interaction (in terms of physical or psychological comfort or uneasiness) and the functional performance and usability (in terms of effectiveness and efficiency). To accomplish this, it must necessarily interact with other disciplines: medicine, cognitive psychology, neurophysiology, psychology of perception. But also with biology, engineering, computer science and ecology, which all transfer important information to the project. Trans-disciplinary research becomes a platform, which is mandatory to catalyze innovative processes, managing contemporary complexity.

Advanced Material Design today is the area of relationship and discussion about different subjects, the basis on which it can give shape to transformations taking place in the field of conscious material design. It is a new scientific perspective which gathers inputs from different disciplines, and exploits unconventional methodologies to create new concepts and material scenarios, responding to a user centred and earth care approach.

Shifting from a macro to micro project scale, and reasoning at the level of objects and systems, Advanced Material Design comes to the definition of material structures and processes. It foresees project solutions mixing up usual material science procedures (made of test, evaluation and characterization processes) with methodologies typical of other fields, while design acts a mediator among different knowledge areas and, using an expression from Denis Santachiara, addresses 'the never devised technological, a free zone between humanistic and technoscientific cultures, in which technology offers itself as the feeling of the artefact and the artificial' (Mendini 1984).

The innovative value of Advanced Material Design consists in the reversal of the traditional problem-solving approach. The material does not exist before, and chosen in the design process, but is born out of the interpretation of the design problem. So, the product concept itself defines the idea of the material, of its texture, performance and behaviour. From here, we continue with experimentation: which structure is suitable to that specific performance? Which the hard and soft characteristics? Which production processes? The material is designed and realized starting from the product concept.

Advanced Material Design is not apart from research of sense in relation to available technological know-how. It adds a research for sense to a merely technical culture, which considers materials as tools for practical object production.

Case 10: Materialecology

Along the path of Advanced Material Design, the research by Neri Oxman stands out. She is an Israeli architect, former medicine student, PhD in design computation, researcher at MIT Media Lab in Boston and winner in 2009 of The Earth Award, a contest created by the environment program of the UN. The goal of her researches is the development of new methodology for the scientific study of material organization, which relies on computer technology to empower and speed up the visualization of material behaviour.
In 2006, Neri Oxman founded Materialecology, based in Cambridge, an interdisciplinary research project which integrates design, computer science, structural engineering, biology and ecology. Researches by Materialecology analyse the behaviour of natural 'living' materials like biological tissues, as well as composite material found in the ground. They simulate the behaviour of these materials through visualization and replication of shapes and their modification in relation to variations of environmental parameters (temperature, humidity, quantity and direction of light, etc.). The goal is to draw reference models from these analysis for the development of objects which behave like living organisms. The design process starts from the project of the material behaviour and structural geometry down to the physical, 3D realization and its structural performance in its vital functions. By underlining the behaviour of materials in relation with the context modification, the researches by Materialecology give sense to the shapes of matter structure. Through the computing of algorithms performing like natural laws, structures are the direct result of assigned parameters, free from formalism and scientifically correct.
Experiments carried out by Materialecology can be regarded as preparation exercises to later projects. Information drew and then processed through the generative design methodology will be used in the design of architectonic structures or objects capable of accurately measuring resistance, elasticity and transparency in order to reduce waist of matter and energy. This is the case of Monocoque (Fig. 11.4), a research finalized to the development of a construction technique based on the concept of structural skin. In the Monocoque prototypes, realized through the poli-jet 3D technology, we find no distinction between structural and non-structural parts. The shape and distribution of materials (solid and empty parts with relative sections) both contribute to structural stability. Another interesting concept is the starting point for Beast project, the chaise lounge built in cooperation with W. Craig Carter, professor of material science and engineering at MIT. As first example of living-synthetic construction, Beast been described 'the shape crystallizing material efficiency'. Material is distributed as a function of the weight of the body the chair holds. The Beast model is inspired by the bones and muscles of human body, and has been realized with resin with a varying resistance watermarked structure printed in 3D. The thickness of structural sections is variable, just

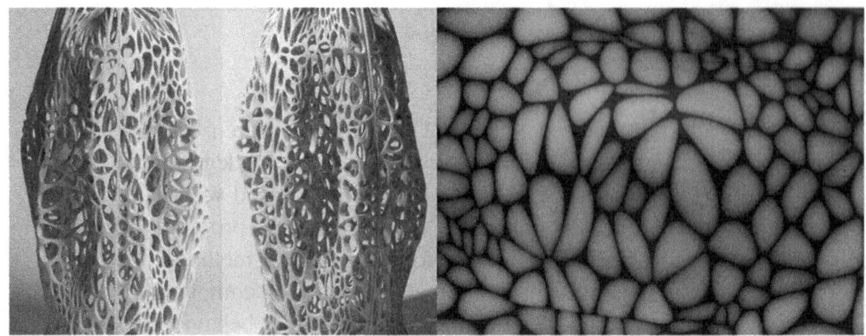

Fig. 11.4 Nery Oxman, Structural Skin Monocoque 1 and Monocoque 2 of Acrylic Composites. Materialecology Cortesy

like flexibility and bending, thus adapting to the different pressures exercised on the chair seat by the different parts on the human body. The varying density of this particular material permits obtaining softer and more flexible zones compared to harder and more rigid ones.

Furthermore, working with Craig Carter and Eugene Bell, a professor in biology and a known pioneer in regenerating medicine, Neri Oxman has developed a new version of solid modeller for fast prototyping, FAB.RE-cology, a device capable of printing objects in which thickness and elasticity can vary in different parts (Fig. 11.5). According to the concept, design can conveniently measure the material and its properties and follow the principles of biology in the realization of objects and architectures.

The idea Oxman promotes with her researches which combine biomimetic vision with design and the construction of built environments, is a scenario where nature is re-integrated inside the artefact, a vision of environmental sustainability of production, driven by material design applying scientific discoveries and technical innovations.

Reference

Oxman, N.: http://web.media.mit.edu/~neri/site/index.html (2011). Accessed 10 Feb 2014

Case 11: Material Beliefs

Material Beliefs is a research group based at Goldsmiths, University of London, in the Department of Design.[5] Currently this group has a core team of five people: Andy Robinson (the project manager), Elio Caccavale, Tobie

[5] Material Belief initial support was provided by the Engineering and Physical Sciences Research Council through a Partnerships for Public Engagement Award (EPSRC grant EP/E035051/1).

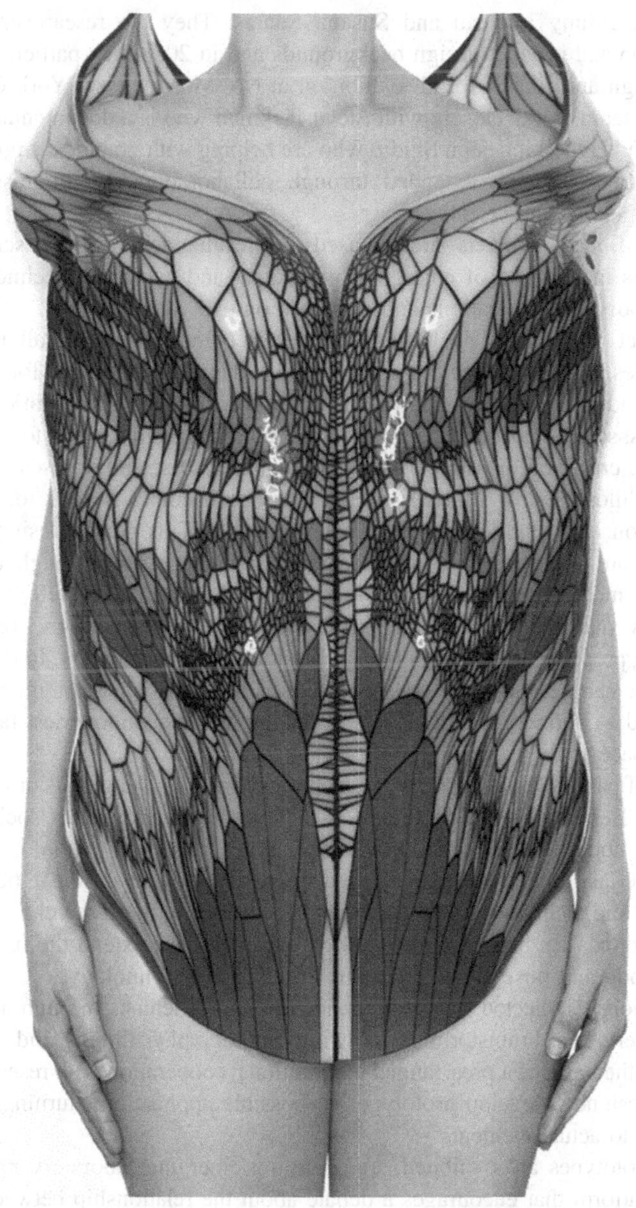

Fig. 11.5 Nery Oxman in collaboration with W. Craig Carter (MIT) and Joe Hicklin (The Mathworks). Arachne is a flexible armour produced with a multi-material printers. Its density and thickness, are informed by the anatomical location of the ribcage. Combined, soft and flexible materials are distributed following continuous web morphology to accommodate for multiple functions, such as protection, enhanced movement, flexibility and comfort

Kerridge, Jimmy Loizeau and Susana Soares. They are researchers with interaction and product design backgrounds and in 2008 they participated in the 'Design and Elastic Mind' exhibition at NY MoMA, New York City.

The core team works closely with Steve Jackman who is a documentary film maker, and Hyperkit design Studio who are helping with graphic design. This initial group is being extended through collaborations in progress with engineers and social scientists.

Material Beliefs projects are focused on results achieved by scientific researches in the field of emerging biomedical and cybernetic technologies and the possible applications and influence on everyday life.

The target of research by Material Belief is complex: first of all to take discoveries out of scientific labs and make them an argument for public debate, and the allow designers and artists work in tight cooperation with engineers, social scientists and even common people from society, to imagine and share creatively possible social consequences of scientific research.

In the philosophy of Material Beliefs design becomes a tool for social exploration of scientific research, and translation of its goals. Design competences are deployed to communicate scientific progress through unconventional methods, though clearer to the general audience, as well as to put ourselves questions, understand the scientists' work methods, research financing procedures, and stimulate discussions about technology. In this way the public can get closer to science, which is normally seen as a monolithic entity and a continuum of practices happening in close laboratories, far from public space.

Instead of passively assuming that a technology may be dangerous or not, the Material Beliefs method leverages design to actively involve society in decisions about materials and processes.

The Material Beliefs method is based on exploration in different fields of scientific and technological research, and focuses on those researches, which reduce boundaries between human body and materials, extend the functions of our body and raise ethics problems in the use of technology.

Information is collected in direct connection with scientists through a set of tools: interviews, brainstorming, drawing, photography, filming and discussion. On the basis of a prearranged explorations, cooperation with researchers allows designers develop prototypes of possible applications, turning information into actual elements.

These prototypes are exhibited, transforming emerging laboratory research into a platform that encourages a debate about the relationship between science and society (Fig. 11.6).

This methods put emphasis with the interaction between the prototypes and statements about social life, rather than the prototypes and business. This method, as well as highlight some new research opportunities for the project design, intend to translate academic knowledge into resources for independent enquiry, to open new way of enabling others to access technology.

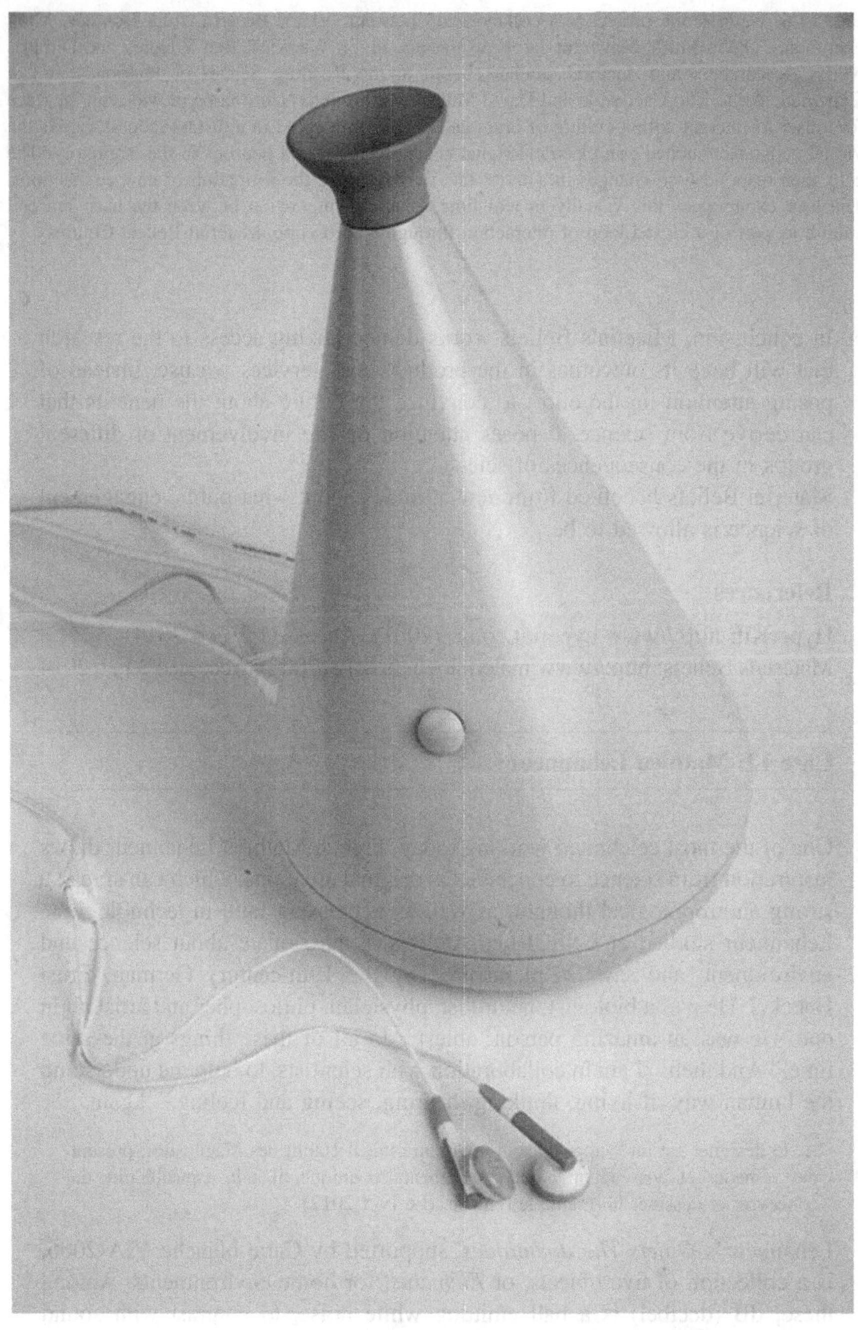

◀ **Fig. 11.6** Neuroscope project is a collaboration between Victor Becerra, Julia Downes, Mark Hammond, David Muth, Slawomir Jaroslaw Nasuto, Kevin Warwick, Ben Whalley and Dimitris Xydas (researchers and doctoral students based at the Reading School of Pharmacy and at Cybernetics) and Elio Caccavale and David Muth. Looking into Neuroscope provides an interface for a user to interact with a culture of brain cells, which are cared in a distant laboratory. As the virtual cells are 'touched', an electrical signal is sent to the actual neurons in the laboratory. The cells then respond with changes in activity that may result in the formation of new connections. The user experiences this visually in real time, enabling interaction between the user and cell culture as part of a closed loop of interaction through Neuroscope. Material Beliefs Courtesy

In conclusion, Materials Beliefs works democratizing access to the research that will have its outcomes in the products and services we use. Instead of posing attention on the effort to convince the public about the benefits that can derive from science, it poses attention on the involvement of different groups in the consequences of science.

Material Beliefs benefited from new attitudes about what public engagement of science is allowed to be.

References

HyperKit: http://www.hyperkit.co.uk (2001). Accessed 2 Feb 2014

Materials Beliefs: http://www.materialbeliefs.com (2009). Accessed 2 Feb 2014

Case 12: Mathieu Lehanneur

One of the most celebrated working today, French Mathieu Lehanneur draws inspiration from science to conceive his original creations, which can spread a strong anthropological thought, as well as expressing faith in technologies.

Lehanneur studied at ENSCI-Les Ateliers, is passionate about science and environment, and says: "I'm inspired by the 19th-century German, Ernst Haeckel. He was a biologist, naturalist, physician, philosopher and artist all in one. He was an amazing person, able to do all of these things at the same time." And then: "I am in collaboration with scientists, to help me understand the human way of living, thinking, hearing, seeing and feeling." Again

> … le designer est un "super docteur". Comme lui, il établit des diagnostics, present des remedes et possède un atout supplémentaire unique, il a la capacité rare de concevoir et dessiner lui-même le remède. (Le Fort 2012)

Lehanneur's *Objets Thérapeutiques*, supported by Carte blanche VIA 2006, is a collection of five objects, or *Éléments*, for home environments. Among these, dB (decibel) is a ball emitting white noise, to contrast with sound pollution. O is a "domestic lung" releasing oxygen thanks to the cultivation of micro-organisms activating when oxygen concentration reduces. These objects are inspired by an early 60s NASA study on health conditions of astronauts showing a high concentration rate of toxic substances in their body due to exhalations by materials (polymers, glass fibers and insulating materials) the space shuttle was made by. A similar volatile substance concentration can easily be detected in our homes, where air can be ten times more polluted that that in urban space, due to furniture products. In fact, wood releases Pentachlorophenol, glues release Formaldehyde, and paint Trichloroethylene, a solvent classified as one of the most carcinogenic substances. These are all substances we can't see but we all breath.

> J'aime travailler sur ce qui ne se voit pas, sur ce qui ne se perçoit pas.

As a designer Matthieu Lehanneur wanted to work on the side-effects, and researching ways of absorbing indoor pollution. I found some plants that were able to do this. The main problem is that they can't do it alone; the air won't go by itself through the soil and roots, the efficient locations for eliminating toxins.

The Andrea air purifier—initially known as Bel Air—is able to increase the efficiency of the plant and help eliminate the toxins.

Mathieu knew that the idea was at risk of appearing rather fanciful, and admitted underestimating the reaction from the general public. He expected that it would take a while for the public to understand his concept and assumed people would believe either it would not work, and accuse the design of being too futuristic, or deny the existence of indoor pollution altogether. Nevertheless, Lehanneur was impressed by the positive reaction the prototype received.

The air we breathe in our homes and at the workplace can be up in collaboration with scientist David Edwards, from Harvard University, Matthieu Lehanneur designed Andrea, an air purifier to counteract the side-effects of traditional design.

In his projects, Matthieu Lahanneur has also explored the universe of medicine, and he has conceived new forms of therapy with the goal of making patients responsible of their own condition, and stimulate the will of medical treatment, like the *Onion medicine*. Everyday the patient takes a layer of the onion until he reaches the kernel, marking the beginning of the healing. This kind of medicine attempts to remove the heaviness and the burden, which often goes with treatments. Pharmaceutical industry has not reacted to this project, but NYC MoMa has listed it in the permanent collection.

In the past, Mathieu would approach them with ideas, inviting and hoping for collaboration, but now he is fortunate enough.

The *Demaine est un autre jour* (Fig. 11.7) project started in the medical field, too. It has been commissioned by the Group Hospitalier Diaconesses Croix Saint-Simon in Paris for the Palliative Care Unit, which takes care of terminal ills.

The projects consists of an 'artificial window' through which one can observe the next day's sky, an electronic device producing a digital simulation of the next day's sky, all the skies in the world, following evolution from dawn to sunset and during the night, thus avoiding a potentially damaging "loop" effect.

Fig. 11.7 Matthieu Lehanneur, Tomorrow Is Another Day. Photo Felipe Ribon

In 2009, doctor Gilbert Desfosses suggested commissioning an artist for a work, which could relieve and stimulate conversation terminal ills and their families during the hard time one lived before death. Thanks to the Gallerie Jérome Poggi, the project was assigned to Lehanneur, who was already known for his therapeutic projects. Thanks to his participation to Transmission Meeting with physicians, nurses, psychologists, Lehanneur's digital window uses the idea climate end time in a very interesting way, with the chance to invite in a conversation about the most common topic, as well as projecting into the future of the day after, or dream and get into a symbolic-emotional dimension not tied to the dramatic situation, and invites us to new forms of liturgy (Ciuffi 2013).

Tomorrow Is Another Day is a synthesis of technology, medicine, science, art and spirituality.

References

Ciuffi, V.: Che cosa sono le nuvole. Abitare **530**, 107–112 (2013)

Le Fort, I.: Le design au service de l'innovation scientifique. We Demain (supplement) **3**, 46–48 (2012)

Lehanneur, M.: http://www.mathieulehanneur.fr (1974). Accessed 2 Nov 2013

References

Bateson, G.: Steps to an Ecology of Mind. University of Chicago Press, Chicago (1972)

Benyus, J.M.: Biomimicry: Innovation Inspired by Nature. Harper Perennial, New York (2002)

Goodsell, D.S.: Fact and fantasy in nanotech imagery. Leonardo **42**(1), 52–57 (2009)

Gorman, M.J.: Future Fictions: What Is… Interview with Anthony Dunne and Fiona Raby. Available via You Tube http://www.youtube.com/watch?v=Zjfp_nmXzyM (2009). Accessed 10 Feb 2014

Kemp, M.: Immagine e verità. Il Saggiatore, Milan (1999)

Latour, B.: Science in action: how to follow scientists and engeneers through society. Harvard University Press, Cambridge (1987)

Maldonado, T.: La speranza progettuale. Einaudi, Turin (1992)

Margolin, V., Margolin, S.: A "social model" of design: Issues of practice and research. Des. Iss. **18**(4), 24–30 (2002)

McDonough, W., Braungart, M.: Cradle to Cradle: remaking the way we make things. North Point Press, New York (2002)

Mendini, A.: Colloquio con Denis Santachiara. Domus **761**, 40 (1984)

Nowotny, H., et al.: The New Production of Knowledge: The Dynamics of Science and Research in Contemporary Societies. Sage, London (1994)

Pasca, V.: Il design nel futuro XXI secolo. In: Treccani Enciclopedia Italiana. Available via http://www.treccani.it/enciclopedia/il-design-nel-futuro_(XXI-Secolo (2009). Accessed 22 Nov 2013

Riccini, R.: Designer e prodotti di fronte al possibile. DIID **38**, 8–21 (2009)

Sterling, B.: Shaping Things. MIT Press, Cambridge (2005)